ALBRECHT KRÜGER

Physics of Solar Continuum Radio Bursts

Physics of Solar Continuum Radio Bursts

by

Dr. rer. nat. habil. Albrecht Krüger

Deutsche Akademie der Wissenschaften zu Berlin
Central Institute for Solar Terrestrial Physics
(Heinrich-Hertz-Institut)

50 Figures, 15 Tables

AKADEMIE-VERLAG · BERLIN
1972

Erschienen im Akademie-Verlag GmbH, 108 Berlin, Leipziger Str. 3–4

Lizenznummer: 202 · 100/588/72
Gesamtherstellung: VEB Druckerei „Thomas Müntzer", 582 Bad Langensalza
Bestellnummer: 5832 · ES 18 D 4, 18 D 3
EDV 761 4333
Printed in German Democratic Republic

Preface

The plan for this book originated from the feeling, that there is a need for a detailed and concise compilation of solar burst physics taking into account the whole complex of the flare-burst phenomenon but being more specialized than current monographs on solar radio astronomy or solar flares. Under these preconditions it seemed appropriate to start from a consideration of the flare-burst problem under the aspect of the interpretation of the continuous radio burst emission which after the author's opinion comprises the central and most important part of the flare-burst problem. The other components of the solar wave and particle emission have been considered so far it was necessary for an understanding of the whole phenomenon.

This book is based on a revised and slightly extended translation of a Doctoral Thesis which gave a critical review of the present knowledge of the physics of solar radio bursts of continuous emission involving the author's work during the past ten years in this field.

The book contains three main parts describing (a) the main observational background, (b) the theoretical basis of continuum radio burst physics, and (c) a synthesis between observations and theory yielding the results and problems of the interpretation of the flare-burst radiations. A bibliography at the end of the book contains more than 400 references.

I whish to express my thanks to Prof. E. A. Lauter, General Secretary of the German Academy of Sciences and Director of the Heinrich-Hertz-Institute, who initiated the edition of this book, for his kind and steady interest in this work. Cordial thanks are also due to Prof. Dr. C. de Jager, Utrecht for his encouragement in writing the text.

The author is further indepted to Prof. Dr. H. Daene, Berlin and Prof. Dr. L. Mollwo, Berlin for their useful advices and help. Finally, it is a pleasure to thank the Akademie-Verlag for the good cooperation at the edition of this book.

Berlin, in December, 1970 — Albrecht Krüger

Contents

List of mathematical symbols and abbrevations

If in the text not otherwise mentioned, most important mathematical symbols are used in the following sense:

A	— sunspot area	(2.3.1.-1)
$A_{E\to E-h_\nu}$	— EINSTEIN coefficient for spontaneous emission	
$A(\gamma')$	— defined by relation	(3.5.4.-6)
$A_\varkappa$	— terms describing the energy supply in the balance equation	(4.2.1.-1)
A_μ	— retarded potentials	(3.4.1.-4)
a_j	$= K_j \cos\theta + \Gamma_j \sin\theta$	(3.4.1.-13)
a_y, a_z, a_θ, a_k	— defined by	(3.4.1.-24)
α	— $n_j \beta \sin\Phi \sin\theta (1 - n_j \beta \cos\Phi \cos\theta)^{-1}$	(3.4.1.-12)
$\boldsymbol{B}$, B_0	— magnetic induction	(4.1.3.-2)
$B(\gamma')$	— defined by	(3.5.4.-9)
B_{js}	— defined by	(3.4.2.-11)
$B_\nu(T)$	— KIRCHHOFF-PLANCK function	(3.2.3.-2)
$B_{E\to E-h\nu}$, $B_{E-h\nu\to E}$	— EINSTEIN coefficiens for stimulated emission and absorption	
$B_\varkappa$	— terms describing energy losses in the balance equation	(4.2.1.-1)
b_j	$= \Gamma_j \cos\theta - K_j \sin\theta$	(3.4.1.-13)
β	$= v/c$	(3.4.1.-2)
β_e	$= v_e/c$	(3.6.1.-4)
β_{th}	$= (3\,k\,T/m_e\,c^2)^{1/2} = v_{\text{th}}/c$	(3.3.1.-3)
C	$= \sqrt{3}\,e^3/(m_e\,c^2)$	(3.8.1.-4)
$C_\varkappa$	$= A_\varkappa - B_\varkappa$	(4.2.2.-2)
c	— vacuum speed of light	
χ	— position angle of the polarization ellipse	(3.3.2.-4)
χ_j	— absorption index	(3.3.4.-1)
χ_{r-s}	— energy difference	(2.1.2.-1)
δ	— attenuation parameter	(2.1.3.-2)
Δs	— difference of ray paths	
Δt	— emission time	(3.5.1.-8)
$\Delta\tau_\nu$	$= \frac{1}{2}(\tau_{\nu,1} - \tau_{\nu,2})$	(3.3.5.-16)
$\Delta\nu_r$	— receiver band width	(3.3.2-.28)

E	— energy	(3.2.2.-6)
E_{tot}	— total energy content	(4.6.1.-2)
$\boldsymbol{E}$	— electrical field strength (components: E_x, E_y, E_z)	
E_{0x}, E_{0y}, E_{0z}	— componentes (amplitudes) of the electrical wave fields	(3.3.2.-3)
e	— electron charge	
$\varepsilon_{\nu,j}$	— emission coefficient	(3.2.1.-1)
	(for isotropic media: ε_ν)	(3.2.3.-1)
η	$= \frac{Y}{(1 - X - i\,Z)} \frac{\sin 2\,\theta}{2 \cos \theta}$	(3.3.2.-15)
η_ν, $\eta_{\nu,j}$	— emissivity	(3.4.1.-18)
η_{ij}	— ergiebigkeit (source function)	(3.4.3.-3)
F	— area on the solar disk	(3.2.3.-13)
F_s	— mechanical energy flux	
$F(E)$	— energy distribution function	(3.2.2.-6)
$f(v)$	— velocity distribution function	(3.1.2.-1)
Φ	— $\sphericalangle$ (v_e, H)	(3.4.1.-7)
φ'	— elevation angle of cylindric coordinates	
φ	— phase angle	(3.3.2.-3)
G_j	— E_z/E_x	(3.3.2.-19)
g	— exponent of injection energy distribution	(4.2.3.-2)
$g(E)$	— statistical weight	(3.2.2.-6)
Γ	— gamma function	
Γ_W	— thermodynamic probability	
Γ_j	$= -i\,G_j$	(3.3.2.-19)
γ	— LORENTZ factor	(3.7.6.-5)
γ'	— exponent of the energy distribution function	(3.5.4.-2)
γ_j	$= (1 + K_j^2)^{-1/2}$	(3.4.1.-12)
γ_w	— ratio of specific heats	(2.1.3.-1)
$\gamma_{s=1}$	— polarization ratio	(3.5.1.-6)
$\boldsymbol{H}$, H	— magnetic field strength	
H_L	— $H \cos \theta$	
H_T	— $H \sin \theta$	
h	— PLANCK's constnat	
h'	— scale height	
I, I_ν	— STOKES parameter, intensity of radiation ($I_{\nu,x}$: linear components of the radiation)	(3.3.2.-5)
I_{ij}	— radiation tensor	(3.4.3.-1)
$I_{\nu,j}$	— intensity of the jth wave mode	(3.2.1.-1)
$I_s(s\,\alpha)$	— BESSEL function of sth order	(3.4.1.-10)
$\boldsymbol{j}$	— current density	
j (index)	$= \begin{cases} 1 \text{ for the extraordinary wave} \\ 2 \text{ for the ordinary wave} \end{cases}$	

K_j	$= i\,R_j^{-1}$	(3.3.2.-14)
k	— BOLTZMANN's constant	(2.1.2.-1)
k'	— (complex) wave number	(2.1.3.-1)
$\boldsymbol{k}$	— wave vector	(3.2.1.-2)
$\varkappa_{\nu,j}$	— absorption coefficient	(3.2.1.-1)
	(isotropic medium: $\varkappa_\nu$)	(3.2.2.-1)
$\varkappa_{\nu,js}$	— gyro-resonance absorption coefficient	(3.4.2.-10)
$\varkappa_{ijkl}$	— absorption tensor	(3.4.3.-3)
L	— characteristic length (dimension) of solar emission regions	(3.2.3.-14)
l	$= \mathrm{sn}_j\,\beta \sin \Phi \sin \theta$	(3.4.2.-1)
λ	— wavelength	
λ_{crit}	$= (4\,\pi^2\,c^3/\omega_p\,\nu_{\mathrm{coll}}\,L)^{1/2}$	(3.3.5.-2)
λ_H	— c/ν_H	(3.4.1.-1)
Λ	— mean free path	(3.7.3.-7)
m	— (rest) mass	
m_e	— electron mass	(3.4.1.-1)
m_π	— proton mass	(3.7.6.-6)
μ	— molecular weight	
μ_j	— real refraction index	(3.3.4.-1)
N	— particle density	(2.1.3.-2)
N_e	— electron density	(3.2.2.-1)
N_i	— ion density	(3.3.1.-7)
N_a	— neutral particle density	(3.3.1.-7)
N_f	— density of fast (relativistic) particles	(4.4.4.-2)
N_0	— density of the undisturbed medium	(4.4.4.-2)
N_B	— average burst number per day	(2.3.3.-1)
N_F	— average flare number per day	(2.3.1.-1)
n_j	— refraction index	(3.3.1.-4)
ν	— observing frequency	
ν_c	— critical frequency for synchrontron emission in vacuum	(3.5.2.-2)
ν_{c,n_j}	— dto. in a medium with refractive index n_j	(3.5.3.-3)
ν_H	— gyro-(cyclotron) frequency	(3.4.1.-1)
ν_s, ν_g	— DOPPLER-shifted gyrofrequency	(3.4.1.-6)
ν_p	— plasma frequency	
ν_{coll}	— collision frequency	
ν_t	— transitional frequency	
ω	$= 2\,\pi\,\nu$	
Ω	— solid angle	
$\Omega_\odot$	— solid angle of the Sun as observed from the earth	(3.2.3.-13)
P_{kin}	— kinetic energy density	(4.4.4.-5)
P_x, P_y	— components of the dielectric polarization vector $\boldsymbol{P}$	(3.3.2.-11)
p	— impuls	(3.7.3.-2)

p_T	— $p \sin \Phi$	
p_L	— $p \cos \Phi$	
p'	— axial ratio of the polarization ellipse $= \text{tg}\, \sigma$	(3.3.2.-4)
Π_j	— wave function	(3.3.3.-1)
Ψ_j	— coupling parameter	(3.3.3.-3)
ψ	$= \sphericalangle (\boldsymbol{k}_j, \boldsymbol{v}_e)$	(3.6.1.-3)
Q	— STOKES parameter	(3.3.2.-4)
Q'	— coupling ratio	(3.3.3.-6)
$Q(E)$	$= A_{E \to E - h\nu}\, h\nu$	(3.2.2.-6)
$Q(E, t)$	$= Q_1(E, t) - Q_2(E, t)$ in	(4.2.1.-2)
q	— electrical charge	(4.1.3.-4)
$\boldsymbol{R}$	— radius vector to the observer	
R_Z	— Zürich definitive sunspot number	(2.3.3.-1)
R_g	— gas constant	(3.3.2.-15)
R_j, $R_{j,s}$	— wave polarization coefficient	(3.3.2.-12)
$R_{\odot ♁}$	— distance Sun-earth	(3.2.4.-3)
r_H	— gyro radius	
ϱ	— degree of polarization	(3.3.2.-2)
ϱ_{ijkl}	— polarization tensor	(3.4.3.-3)
ϱ_0	— neutral gas density	
S_ν	— flux density	(3.2.3.-12)
$\boldsymbol{S}$	— POYNTING vector	(3.4.1.-11)
s	— ray path	(3.2.1.-1)
s	— order of harmonics	(3.4.1.-5)
σ	— electrical conductivity	(3.7.1.-1)
σ'	— heat conductivity	(2.3.-1)
T	— temperature (thermal equilibrium)	
T_e	— effective temperature	(3.2.3.-13)
T_s	— radiation (brightness) temperature	(3.2.3.-8)
t	— time	
t_d	— dissipation time	(3.7.1.-1)
t_D	— escape time for relativistic electrons from a magnetoplasma	(3.8.2.-4)
t_e	— energy exchange time	(3.8.2.-2)
t_r	— half-power time	(4.1.2.-8)
θ	— $\sphericalangle (\boldsymbol{k}, \boldsymbol{H})$	(3.4.1.-8)
ϑ	— $\sphericalangle (\boldsymbol{k}, \boldsymbol{v}_{\text{gr}})$	(3.2.1.-2)
ϑ_j	— $\sphericalangle (\boldsymbol{k}_j, \boldsymbol{v}_{\text{gr}})$	
τ	— relaxation time	
τ_f, τ_s	— collision times for fast electrons and the ambient medium, respectively	(4.4.4.-2)
$\tau_{\nu,j}$	— optical depth	(3.2.1.-3)
U	— STOKES parameter	(3.3.2.-4)
V	— STOKES parameter	(3.3.2.-4)

V'	— volumen	
$\boldsymbol{v}$	— velocity (with components v_x, v_y, v_z)	
v	— electron speed	
v_{gr}	— group velocity	
v_k, v_{k_j}	— phase velocity	
v_m	— ALFVÉN velocity	(3.7.3.-8)
W	— (kinetic) energy	(3.7.3.-4)
W_s	— energy emitted at the s th harmonic	(3.4.1.-11)
W_ν	— radiated energy per frequency interval	(3.5.2.-1)
w	— velocity	
w_T, w_L	— velocity components with respect to the magnetic field direction	(4.1.3.-2)
w_D	— drift velocity	(4.1.3.-2)
X	$= \omega_p^2/\omega^2$	
x, y, z	— Cartesian coordinates ($x \| \| H_T$, $z \| \| k_j$)	
ξ	— defined by	(3.3.5.-7)
Y	$= \omega_H/\omega$	
Y_T	— $Y \cos\theta$	
Y_L	$= Y \sin\theta$	
Z	$= \nu_{\text{coll}}/\omega$	
Z_0	$\triangleq$ critical collision frequency	(3.3.2.-16)
Z_k	— atomic number	(3.2.2.-1)
z_{js}	— defined by	(3.4.2.-11)
ζ	$= (Y \sin^2\theta)(2\cos\theta)^{-1}$	(3.5.1.-5)

Bold-face letters denote vectors, overlined quantities — (spatial) mean values, stars — conjugated complex values. If no other remark is given, cgs units are used.

1. Introduction

During the past twenty-five years solar radio astronomy became an extended field of research which is closely related to many branchs of sciences. The whole subject already covers special monographs (KUNDU, 1964 and 1965; ZHELEZNYAKOV, 1964 and 1970) and the importance of solar radio astronomy is reflected by the work carried out in this field in more and more countries.

At present most difficult and interesting problems are connected with the investigation of the flare-burst phenomenon. The instationary physical processes causing solar flares and different types of radio bursts are typically associated with radiations in other parts of the electromagnetic spectrum as well as with the emission of particles of different energies. For this reason, besides the general importance for astrophysics, the investigation of the flare-burst complex is intimately related to basic aspects of solar-terrestrial physics, space research and geophysics.

Solar burst physics is to a great extent applied plasma physics containing a number of open questions which are only hoped to be solved by the joint efforts in the fields of theoretical analysis, laboratory experiments and solar observations.

Stimulated by the great projects of the International Geophysical Year 1957/58 (IGY), the International Geophysical Cooperation 1959 (IGC), and the International Years of the Quiet Sun 1964/65 (IQSY) a growing number of solar radio observations have been made at several stations around the earth and became available more or less regularily until now.

As a first step, these observations have been treated phenomenologically by means of studies of special burst events or statistical investigations basing on a greater number of cases. In a later stage, attempts of a physical interpretation of the observed phenomena followed. Recently, the above mentioned international projects have been replaced by new international cooperative programs which now are concentrated on the solution of special, very concrete problems combining the results, obtained by application of a greater number of different methods. An example of this kind of work is given by the Proton Flare Project (PFP) which is included in the projects of the Inter-Union Commission on Solar-Terrestrial Physics (IUCSTP).

But not only the observations, also the theoretical tools necessary for a treatment of burst physics have been developed in several directions during the recent years. In this way the theorey of bremsstrahlung and gyro-synchrotron radiation is available in the framework of the single particle approximation and was completed by taking into account the effects of the plasma medium. Also the wave propagation characteristics of a magnetoplasma are wellknown. With this respect, many of the burst radiation properties can be explained. But it must be stated, that the problem of the origin of the particle accelleration leading to the flare-burst event remains still unsolved.

Concerning the different types of solar radio bursts, especially those which exhibit a continuous emission reveal a great importance for solar physics and solar-terrestrial physics, because all big solar events associated with geophysical effects are characteristically connected with continuous radio burst emissions containing essential informations about the physical processes originating in the solar atmosphere.

As a "continuous" radiation in the radio frequency region a broad-band radiation will be defined in contrast to the narrow-frequency emissions of the drift bursts. In most cases the spectral extent of a burst emission is correlated with its extent on the time axis, i.e. pips of a short duration are not regarded as continuous emission. It is obvious that the above given convention a priori is without reference to a certain emission mechanism. It should be further noted, that this definition of "continuous" radiation in the radio region has nothing to do with definition of continuous radiation commonly used e.g. in the optical spectral range, because line emissions up to now do not play any noticable role in the solar radio spectrum.

Adopting the above given definition of continuous solar radio burst emission, we have mainly to consider all types of bursts in the microwave region including the mm-, cm-, and lower dm-range, and the complex phenomenon of type IV bursts. The explanation of the deeper origin of these emissions seems to be one of the central points necessary for an understanding of the whole flare-burst problem.

It should be noted, that the phenomenologically continuum-like emission of the type V bursts is believed to have really a non-continuous character similar to that of the type III bursts and therefore is out of the scope of the present consideration.

2. Basic phenomenological results

2.1. Physical conditions in the undisturbed solar atmosphere and in active regions

In this chapter some fundamental observational facts are compiled which give the phenomenological-statistical background for a study of burst physics. It must be kept into mind, that for this purpose not only radio data but also optical and other observations must be taken into consideration in order to obtain a complete as possible picture of the interesting phenomena. For this reason in the following the undisturbed and disturbed physical conditions outside and inside of the centres of activity, respectively, are briefly re-examined.

2.1.1. Photospheric properties

The photosphere can be regarded as the lower boundary of that part of the Sun (i.e. the solar atmosphere) which is of direct interest for radio astronomy. Only mm- and submm-waves have a chance to originate in levels near the photosphere. Besides that, the photosphere has a great influence on the higher levels of the solar atmosphere which are important for the generation of radio waves.

The photosphere is the at present best investigated part of the Sun. The greatest part of the Sun's undisturbed electromagnetic radiation (optical light) comes from the photosphere. Physically the photosphere is characterized by a steep gradient of particle density and a minimum of temperature ($T_{\text{kin}} = 4500$ °K). In a layer of a thickness of about 400 km the optical depth rises by two orders of magnitude, whereas most of the radiation comes from a layer of 100 km thickness. The accurate position of the minimum temperature in the upper photosphere could be located by means of rocket observations at wavelengths between 1000 and 2000 Å (Tousey, 1963).

In the photosphere the radiation equilibrium is predominant. Nevertheless differences between model computations based on the assumption of radiation equilibrium and the observed temperatures as a function of the optical depth indicate a certain influence of convection (de Jager, 1965). The source of the convection consists in the subphotospheric hydrogen convection zone. In this zone a flux of mechanical energy originates which comes up to the chromosphere and corona and has an essential influence on the structure in these levels.

The gravitational acceleration in the level of the photosphere amounts to $2.74 \cdot 10^4$ cm s^{-2} and the centrifugal acceleration at the equator is -0.587 cm s^{-2}. The rotational energy of the Sun amounts to $2.5 \cdot 10^{42}$ erg. The minimum speed of particles escaping from the Sun's gravitational force is $6.177 \cdot 10^2$ km s^{-1}.

2.1.2. The chromosphere

The chromosphere which covers the photosphere up to a height of $5 \cdot 10^3$ to 10^4 km is already of direct interest for radio astronomy. The chromosphere can be regarded as the transition layer between the relatively cool neutral gas of the photosphere and the hot and thin plasma of the corona. Inside the chromosphere the temperature raises from about 4000° to about 40000 °K. In the same interval the density diminishes by about five orders of magnitude. Accordingly the degree of ionization increases with growing height.

The variety of spatial structural details is one of the most remarkable features of the chromosphere and points to the action of different forms of plasma motion and turbulence. Thus, it can be stated that a "quiet" homogeneous chromosphere does not exist. For an interpretation of optical, e.g. spectrohelioscopic, measurements it must be taken into account that due to inhomogeneities local differences of the excitation temperature take effect in the structure of the cores of the FRAUNHOFER lines. The changes of the excitation temperatures follow, in a first order, BOLTZMANN's formula

$$\frac{N(E_r)/g(E_r)}{N(E_s)/g(E_s)} = \exp\left(\frac{\chi_{r-s}}{kT_{r-s}}\right). \qquad (2.1.2.\text{-}1)$$

Such local differences of temperature and density arise e.g. in the presence of local magnetic fields. Bu , on the other hand, it is indicated, that the excitation and ionization temperatures of the chromosphere are different from the kinetic temperatures of the chromosphere determined by the energy distribution of the particles. Therefore differences from the local thermodynamic equilibrium are noticeable.

Because radio observations scarcely give informations about the fine structure of the emitting regions, one is thrown on the consideration of the results of observations at much shorter wavelengths, e.g. in the optical region.

Three main types of chromospheric structural elements are to be mentioned: The fine mottling, the coarse mottles and the coarse network.

The fine mottling is the most common structural feature of the high chromosphere (5000—10000 km above the photosphere). From an analysis of line spectra temperatures between 1000 and 2000 °K can be derived, i.e. the fine mottling (spiculae) consists of cold and dense elements (about 500—1000 km diameter) which are surrounded by coronal matter. A similar distribution of cold and hot elements is also present in the lower chromosphere. Here the temperatures are between 4000 and 10000 °K. But the one-to-one connection of these hot and cold elements in the low and high chromosphere appears to be not yet fully recognized.

Besides the fine mottling there are other structure elements in form of the coarse mottles of 2000 to 6000 km diameter, which are best seen in spectroheliograms of the medium chromosphere.

Finally, large scale structures especially in the lower chromosphere are present in form of the coarse network. This is closely connected with the structures of the magnetic field named supergranulation by LEIGHTON et al. (1962). Inside the cells of the network horizontal motions of 0.3—0.5 km s^{-1} are present, while at the boundary of the cells (dark mottles, rosette structures) a stream of matter with an average speed of 1.5—2.0 km s^{-1} is directed downwards (cf. DE JAGER, 1965).

On the basis of optical, ultraviolet and radio observations a number of chromospheric models have been proposed, but as a matter of fact, no one is satisfactory in all respects. Some rough and schematic data are compiled in fig. 1.

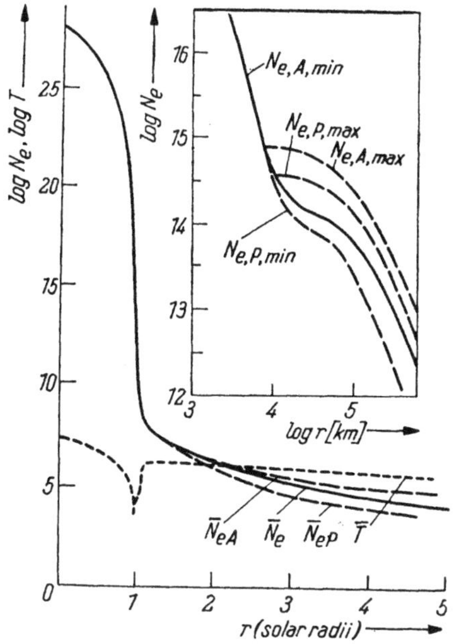

Fig. 1. Average radial distributions of temperature and electron density in the solar atmosphere (after ALLEN, 1955; DE JAGER, 1961)
A — equator, P — pole

2.1.3. The Corona

The corona is not only by far the most extended part of the solar atmosphere but also the most interesting part of the Sun for radio astronomy. The transition from the chromosphere to the corona is characterized by the steep gradient of temperature due to a vast increase of the dissipation of mechanical waves coming from the hydrogen convection layer.

Optical observations of the coronal radiation exhibit three main components, the F-corona, the K-corona, and the L-corona. The white light of the corona resulting from the K- and F-corona exceeds the L-corona (i.e. line-emission) by a factor of hundred. The K-corona as a pure continuous emission with a high degree of polarization, consists of photospheric light scattered at free electrons. In comparison to this, the F-corona arises from scattered light at interplanetary dust particles. The spectrum of the F-corona exhibits FRAUNHOFER lines and is especially in the outer parts of the corona predominating (in a distance of two solar radii from the centre of the sun F- and K-corona are nearly of the same intensity.) The F-corona has its continuation in the zodiacal light. Observations of occultations of cosmic radio sources by the solar corona at distances up to about hundred solar radii from the Sun's centre lead to the conclusion that the corona is not symmetrically even in the equatorial plane. A flattening of the corona towards the apex of the motion of the Sun in space is to be noticed.

The corona is not so rich of structural details as it is characteristical for the chromosphere. Typical structures of the corona seen in optical light are merely the streamers and polar plumes. The coronal streamers have been interpreted as magnetic tubes coming from the photosphere and supplied with matter from below. It must be mentioned that fast macroscopic motions are not observed under undisturbed coronal conditions. Nevertheless, from the existence of the solar wind it is known, that there is a permanent outflow of matter from the Sun, so that without supply from below after a comparatively short time the corona would be vanish into space.

Investigations of the solar X-ray emission are of particular interest for a completion of the studies of solar radio emission. Similar as the solar radio emission, also the X-ray emission in first order is proportional to N_e^2. In the corona the electron density decreases in outward direction with a scale height of about $5 \cdot 10^4$ km. Accordingly, under quiet conditions the X-ray emission decreases with a scale height of $2.5 \cdot 10^5$ km. But the X-ray spectrum as well as the radio waves are strongly influenced by solar activity as it will be discussed in the following sections.

The explanation of the high values of corona temperature and the sharp transition between the chromosphere and corona are of general interest not only for studies of the quiet corona itself but also for other cases of generation of high temperature plasmas, e.g. special burst sources. In the following some remarks on this question are given.

For the flux of mechanical radiation coming from the hydrogen convective zone a strong dissipation of energy is expected, if the Mach number becomes larger than unity. This dissipation occurs at relaxation paths of the order of the scale height $h' = \frac{1}{d/dr\,(\ln N)}$. The accoustic energy flux is proportional to $\varrho_0 A^2 \omega^2 v_s$ where ϱ_0 — density, A — amplitude, ω — frequency, and v_s — group velocity of the acoustic waves, leading to a dissipation of the order of $(\varrho_0 A^2 \omega^2 v_s)/h'$. But only if the amount of kinetic energy dissipated by the shock wave is greater than the energy losses which are mainly caused by radiation, heat conductivity and convective processes, a considerable increase of the temperature will be noted.

A comparison between the energy dissipation and the quantum of electromagnetic radiation (mainly free-bound transitions) leads to the conclusions that, as a consequence of the proportionality of the intensity to $N_e^2 T^{-1/2}$, above a certain critical density practically all energy can be converted into radiation. Otherwise, below this critical density, a heating of the medium takes place (UNSÖLD, 1960). This behaviour probably could provide an explanation of the sharp boundary between chromosphere and corona as well of the approximate proportionality between the mechanical energy flux and the density of the chromosphere. Due to the dissipation processes (magnitude approximately 10^5 erg cm^{-2} s^{-1}) after a certain relaxation path the mechanical energy flux is decelerated. As a consequence, large gradients of temperature arise and the action of heat conduction becomes important (ALFVÉN, 1941).

The influence of heat conduction was considered e.g by STURROCK (1964a). Basing on the dispersion relation for acoustic waves in a medium with the heat conductivity σ

$$\omega\left(\omega^2 - \frac{\gamma_\omega kT}{m} k'^2\right) + i\frac{2}{3}\frac{\sigma' k'^2}{k N}\left(\omega^2 - \frac{kT}{m} k'^2\right) = 0 \qquad (2.1.3.\text{-}1)$$

(k — BOLTZMANN's constant, k' — complex wave number) the energy density can be estimated. A maximum is obtained, if the damping parameter

$$\delta = \frac{2}{3}\frac{\sigma' k'^2}{k N} \qquad (2.1.3.\text{-}2)$$

is about unity (cf. fig. 2).

Concerning the influence of convection it should be mentioned, that this influence decreases with increasing height above the photosphere. Therefore convection appears not to play an important role in the solar corona. Nevertheless, the characteristic spiculae structure of the transition from the chromosphere to the corona could be explained under assumption of the action of convective motions.

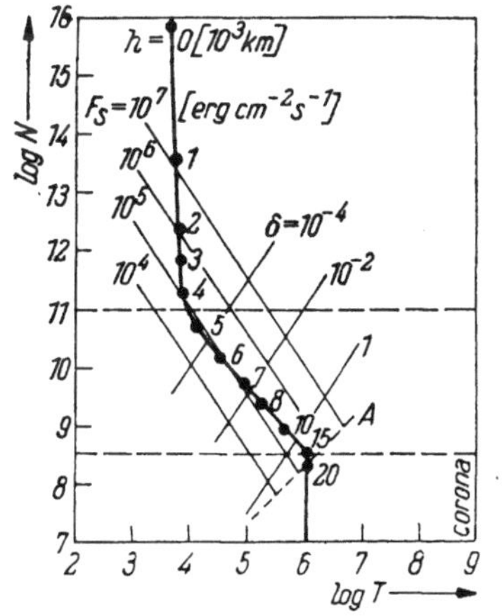

Fig. 2. Density-temperature profile in the solar atmosphere. Straight lines denote constant energy flux of acoustic waves ($M = 1$) and constant damping by thermal conduction, respectively (after STURROCK, 1964a)

2.1.4. Solar active regions

The centres of solar activity are surrounded by the "quiet" solar atmosphere. Though not important for the total energy outflow from the Sun as expressed by the solar constant, the phenomena of solar activity influence strongly the picture of the Sun obtained in different spectral regions, so that an ideal Quiet Sun even under conditions of the solar minimum appears only as a useful fiction. Moreover, almost all phenomena of solar-terrestrial relationships depend in their origin on the complex of solar activity.

A centre of solar activity is in its structure three dimensional covering all heights and levels of the solar atmosphere. In this way solar activity appears as the totality of numerous single phenomena which are more or less closely related to the appearance of variable magnetic fields most evidently demonstrated by the occurrence of sunspots.

A compilation of the most important characteristics of solar activity is shown in table 1, where the single phenomena are plotted in dependence of their height of origin in the solar atmosphere and their characteristic lifetime.

Undoubtedly the knowledge of the concrete physical conditions being present in solar active regions is essential for a correct interpretation of solar radio phenomena. On the other side, the results obtained by radio astronomical methods contribute to the whole picture of solar activity, in which the radio phenomena represent only special components.

Tab. 1. Main phenomena of solar activity

Process / Level of origin	quasi-stationary	instationary
Photosphere	Spots	
	Plages (faculae)	White light flares moustaches bright points
Chromosphere Corona	Coronal condensations (L-corona) S-component of X- and radio emission quiescent and spot prominences enhanced solar wind streamers and R-ray	Flares microwave bursts X- and γ-ray bursts eruptive prominences radio bursts of different spectral types cosmic and subcosmic radiations

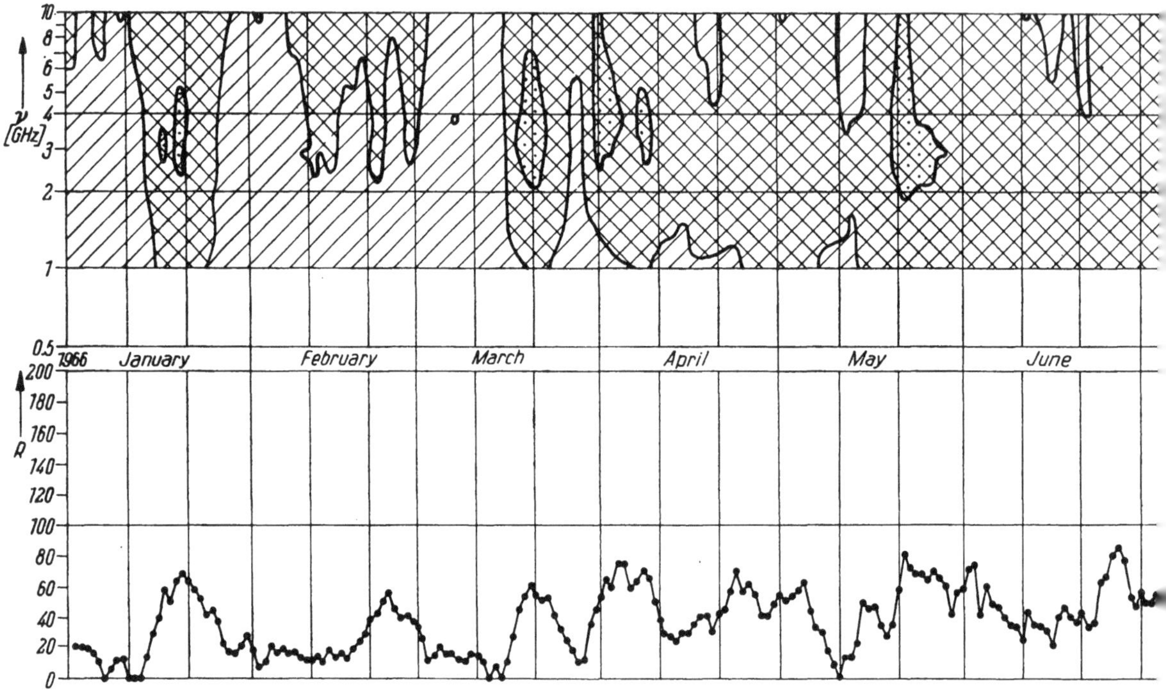

Fig. 3. Dynamical spectrum of the *S*-comp

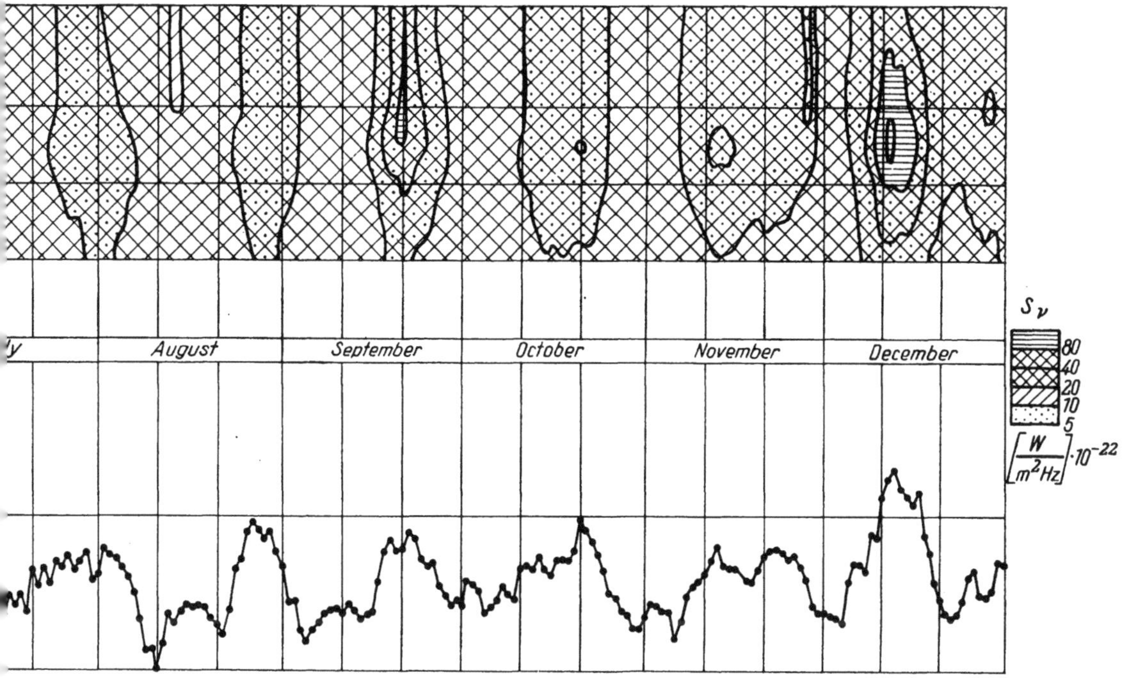

ıd sunspot numbers during 1966

With respect to slow changing solar radio phenomena the formation of permanent coronal condensations is one of the most interesting points. The development of coronal condensations starting already before the appearance of spots and pores is reflected by local enhancements of electron density and magnetic fields. The existence of coronal condensations is indicated by emissions of coronal lines (e.g. 5303 Å), the appearance of the continuous emission of the slowly varying component (S-component), and also by enhanced radiations in the ultra-violet (e.g. Lyman-α) and X-ray (5—60 Å) regions.

The spectrum of the S-component (as well as of the X-radiation too) reveals interdiurnal changes according to the time scales of the development of the active regions, and to the centre limb variation of the radiation caused by the solar rotation (cf. fig. 3). The latter causes a 27-day period which is clearly detected by means of autocorrelation studies. For the X-ray emission also shorter periods due to characteristic lifetimes are well expressed (cf. e.g. Krüger et al., 1969).

From the point of view of solar radio burst physics an examination of the S-component is of special interest in the following two directions. Firstly, it is expected that the majority of at least microwave bursts takes its origin typically in the source regions of the S-component. Secondly, the continuous emission of the S-component delivers model cases of the generation of an enhanced solar radio emission. In this way some results of the theory of bremsstrahlung and gyro-resonance emission, which are hitherto succesfully applied to the interpretation of the S-component, in a modified form are partly applicable to certain questions of burst emission.

On the other side, the large differences between the nature of the S-component and of some burst phenomena, even in the microwave region cannot be ignored. Such differences exist with regard to the characteristic lifetimes of the source regions and the less variable average spectrum of the S-component.

As it is well known, there is a close correlation between the flux of the S-component and the number of sunspots or the plage area. This fact indicates strongly that in an active region the different height levels are to be considered in a close interaction with each other.

The at first and best observed phenomena of solar activity are the sunspots. The dimensions of the umbrae vary between about 2000 and 20000 km. Typical diameter of the penumbrae are 5000 to 50000 km. The observed magnetic field strengths in the centres of the umbrae are ranging from about 150 to 4500 gauss. These magnetic fields represent a mighty reservoir of magnetic energy ($\leqq 10^{34}$ ergs). Besides this, laminar motions of matter along the field lines (Evershed effect) represent a smaller content of kinetic energy.

For the interpretation of the polarized radio emission (e.g. type IV bursts) the knowledge of the distribution of magnetic polarities in a active region is needed. In the past solar cycle Nr. 19 (1954—1964) the leading (p-) spots were connected with northern polarities of the magnetic field on the northern solar hemisphere (cf. table 2).

Tab. 2. Interpretation of polarized radio emission by the magneto-ionic wave modes according to the leading spot hypothesis

Solar cycle No.	19 (uneven)		20 (even)	
Handedness of polarization	right	left	right	left
Source on northern	$e(x)$	o	o	$e(x)$
southern solar hemisphere	o	$e(x)$	$e(x)$	o

Other long living phenomena of solar activity as prominences up to now did not show any remarkable direct relation to special characteristics of the radio emission.

Concerning further expressions of slowly varying solar activity, the observations of an increased solar wind from active regions (R-rays) should be mentioned. The R-centres can be considered as the continuation of the active regions into the outer corona, they are responsible for the generation of recurrent geomagnetic storms. They are not to be confound with the usual coronal rays which are extending only to 0.5 astronomical unit distance from the Sun being observed often associated with stationary filaments corresponding to a late phase of the development of an active region.

Considering the characteristics of solar activity during the eleven-year cycle, some not yet fully explained pecularities can be stated. It must be noted that different phenomena exhibit different periods in the solar cycle. A striking feature is the existence of two clearly distinct maxima for phenomena connected with a high energy exchange (proton flares) or taking place high in the corona. Fig. 4 demonstrates this behaviour schematically for some indices of solar activity. The occurrence of these two maxima of the solar cycle may be connected with the splitting of the butterfly diagram into two patches at different heliographic latitudes found by GNEVYSHEV (1963).

2.2. Types of solar radio bursts

After consideration of the long-term characteristics of the solar activity now we are dealing with the basic phenomenological properties of the short-term changes, i.e. the flare phenomenon. At first we start with a discussion of the low frequency edge of the solar electromagnetic spectrum, i.e. the range of radio waves.

It is not our intention to give here a complete description of all observational facts and details which would require much more space. Besides, such a work has been done already at other places (cf. e.g. FOKKER, 1960; KUNDU, 1965;

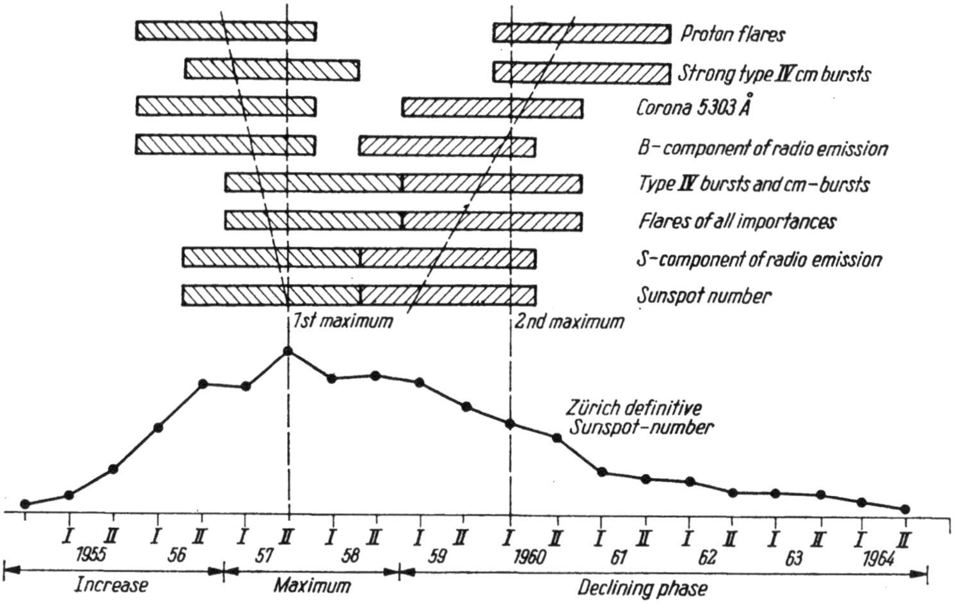

Fig. 4. Schematic representation nof the time characteristics of different phenomena of solar activity in the eleven-year solar cycle

KRÜGER, 1961; WILD et al., 1963, MAXWELL, 1965). It is rather the aim here to point out some fundamental properties of the special types of solar burst radiation which appear as essential for the interpretation of solar continuous radio burst emission when all known phenomena of the whole flare-burst complex are taken into account.

A compilation of the basic characteristics of the different burst types is presented in table 3 which provides some general orientations about the multitude of solar radio burst emission.

2.2.1. *Radio bursts in the cm-region*

Besides the type IV burst continua in the decimetre-, metre-, and decametre-regions, the continuous burst emission of the centimetre-region (microwave bursts) is one of the main objects of the present treatise. Examples of most typical morphological time profiles of microwave burst fluxes recorded at single frequencies are shown in fig. 5. According to such time profiles the following basic morphological types of microwave bursts can be distinguished:

— gradual bursts
— impulsive bursts
— complex bursts
— type IV bursts.

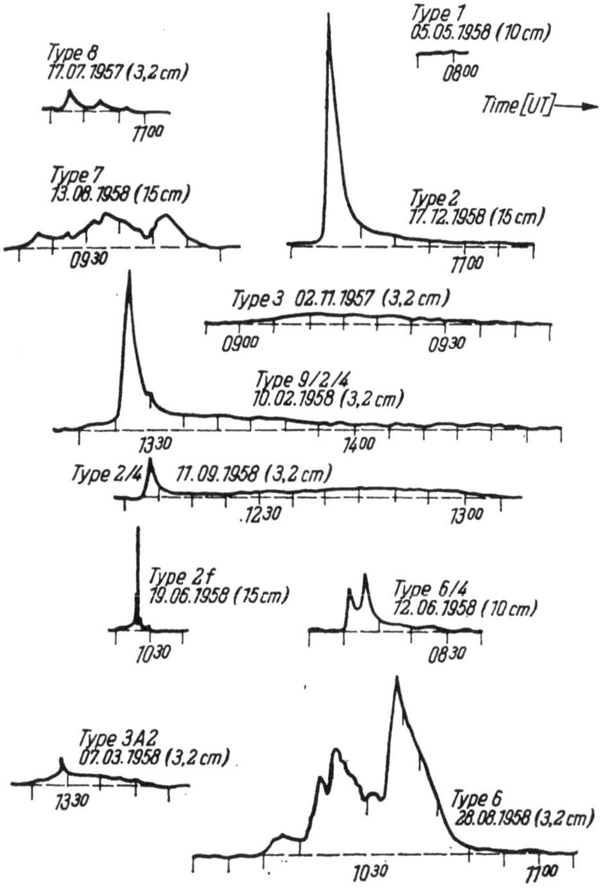

Fig. 5. Characteristic morphological forms of solar microwave bursts. The types refer to an extended Covington-classification. (Wavelengths are given in parentheses)

It must be noted, that the boundaries between these types are not sharply marked and many transitions from one type to another are quite possible (cf. section 2.5). Therefore it is not surprising, that various different classifications have been introduced and used by different authors (Kundu, 1959, Covington, 1951; Krüger, 1962b; Wild et al., 1963). Here the classification proposed by Wild, Smerd and Weiss (1963) has been adopted, which fits best the main current ideas on an physical interpretation of the burst properties, though the nature of the gradual bursts remains not yet clear.

There arises the question, whether essential physical differences between the bursts of the types post-burst increase and gradual rise and fall exist or not. The latter type, however, possibly includes emissions of different character depending of the association with type IV burst components in other spectral

Tab. 3. Compilat

Type	"Microwave-bursts" Small effects (impulsive)	Gradual bursts	μ	Type IV dm	mA
Frequency range	$\gtrsim$ 1000 MHz			150–1500 MHz	$\lesssim$ 200 M
Emission	continuous				
Harmonics	not observed				
Variability	weak	very weak	weak	fine structure	weak
Number in solar maximum	$\lesssim$ 1 per hour	$\lesssim$ 1 per day	~1 per month, very big events are avoiding the solar maximum ?		
Bandwidth [MHz]	500–10000		2000–6000	~1000	~200
Frequency drift	no drift				sometim similar type II
Duration	<10 min	<10 min	10 min . . . several hours		
Max. flux $\frac{W}{m^2\,Hz} \cdot 10^{-22}$	$\lesssim 10^3$	$\lesssim 10^2$	$\lesssim 10^4$	$\lesssim 10^5$	$<10^5$
Equivalent temperatur °K	$\sim 10^6 - 10^{10}$	$\lesssim 10^6$	$\lesssim 10^9$		$10^7 - 10$
Circ. polarization	no or partial		partial	partial . . . strong	weak
Mode	extraordinary	?	extra-ordinary	ordinary	extra-ordinar
Lin. polarization	partial	?	partial		$\lesssim$10%
Directivity	moderate		weak	moderate	
Height of the Source	$-0{,}05\ R_\odot$			$0{,}1-0{,}4\ R_\odot$	0,5–5
Diameter	1–4′	<1′ 2	2–4′	3–5′	6–12′
Movement	stationary				ascendi ($\sim 10^3$
Emission mechanism	bremsstrahlung + cyclotron emission	bremsstrahlung (+ cyclotron emission)	synchrotron emission		
Origin	thermal and superthermal electrons		in magnetic fields trapped electron		
Associated phenomena	flares, SID's (X-rays)		flares, proton streams, aurora, geomagnetic storms etc.		
Pecularities	fluctuations, groups, homology	superposed single bursts	often two phases	absorption effects	"spagh structu

burst characteristics

"Meter-bursts"					
	Noise storms				
m B	continuum	Type I	Type II	Type III	Type V
≲200 MHz	≲5000 MHz		20 – 200 MHz	≲500 MHz	≲200 MHz
		non-continuous			cont. ?
		second harmonic			?
…owly varying		strong	complex	group structure	weak
	1 – 3 per week		∼1 per 50 50 hours	≲ 3 per hour	≲ 1 per hour
≲200		∼4	∼5	50 – 100	
	no drift	sometimes similar type III	< 1 MHz/sec	∼20 MHz/sec	(following type III)
…everal …ours	hours . . . days	<1 sec (groups ∼0.5 min)	∼5 – 30 min	≲ 10 s (groups ∼1 min)	≲1 min
$\lesssim 10^4$		$\gtrsim 10^4$	$< 10^5$		
$\lesssim 10^{10}$		$\gtrsim 10^{10}$	$\sim 10^8 - 10^{11}$		
strong			very weak	partial	
ordinary			—	?	?
≲10%	?	very weak		sometimes observable	
strong			weak		
$0{,}2 - 0{,}1\ R_\odot$			$0{,}2 - 1\ R_\odot$	$0{,}2 - 2\ R_\odot$	
…– 6′	2 – 10′	1 – 6′	6 – 12′	3 – 12′	
stationary			assending $\left(\sim 10^3 \frac{\text{km}}{\text{s}}\right)$	ascending (descending)	stationary ? ?
probably (…erenkov-)plasma waves		plasma waves $L - T$ interaction			?
		?	shock waves	electron streams	
	sometimes flares		Type IV	often flares (flash-phase)	Type III
		reversing drift-pairs	frequency splitting, her-ring-bone structure	*U*-bursts, intermediate drift bursts	

ranges or not. Another experimentally not yet solved problem consist in the question of continuous transition between the phenomena of gradual bursts and the S-component concerning their life times. Recently rises of intensity without immediate subsequent fall of intensity could be detected (Covington, 1969).

Effects of post-burst decrease of radiation are also of special interest, which can be interpreted by absorption effects in certain levels above the burst source caused by ejected matter from the burst region (Yurovsky, 1969).

In the millimetre region no qualitative new features of the burst morphology are observed, so that the classification of the microwave bursts can be adopted accordingly also for this region.

While the burst morphology reflects well the dynamics of the burst processes, the burst spectrum contains other important informations which for the solution of the question of the emission mechanisms and the estimation of solar plasma parameters are even more important than the burst morphology. The following main types of burst spectra can be distinguished in the microwave region with respect to the position of spectra maxima and the gradients of the spectral curves in that region:

a) Spectra increasing monotonously towards higher frequencies or ending in a horizontal branch (mm-bursts),
b) spectra increasing towards lower frequencies (dm-bursts),
c) spectra showing a maximum within the centimetre-region (cm-bursts).

Of course, also combinations of the above mentioned basic types as well as changes during the lifetime of a burst event are possible.

In a first approximation the position of the burst emission on the frequency axis is determined by the magnetic field and the height of the burst source in the solar atmosphere. The energy distribution of the emitting particles and the spatial extension of the burst source are expressed by the frequency extent of the burst emission. With this respect, the solar microwave burst spectra can be also roughly divided into

A) broad band continua, and
B) spectra of a relatively small frequency extent (less than, say, 1000 MHz).

Furthermore measurements of the polarization of the burst radiation are of interest especially for an examination of the emission mechanism and the conditions of wave propagation as discussed later. Typically the degree of polarization of solar microwave burst radiation ranges between 30 and 60 per cent whereas the amount of circular polarization is always dominating. For a satisfactory determination of the polarization effects, however, the consideration of fine structures is required.

By this and other reasons measurements of a high angular resolution are of a great value but also expensive and therefore, unfortunately, only in a restricted number available. The best achieved resolution at the time being is about 1 minute of arc, which is still insufficient to resolve fine structures of the burst sources. Some results of these measurements are included in table 3.

A special remark should be given concerning the problem of type IV bursts. As general accepted since the Kyoto Conference on Earth Storm and Cosmic Rays and the IAU Symposium No. 16 on the Solar Corona in 1963 the definition of the type IV bursts has been extended to the cm-region. Thus the strongest complex and great bursts in the microwave region are to be considered as type IV cm bursts. It should be noted that the definition of type IV bursts as strong long-period continuous radiation primarily was based on morphological burst characteristics. Therefore, without a founded physical conception, there are more quantitative than qualitative differences between the usual complex microwave bursts and the type IVcm bursts. For this reason in some cases the practice of classification prove to be difficult. Therefore an improved classification scheme basing on physical criteria seems to be desirable.

2.2.2. *Decimetre bursts*

Decimetre bursts represent the transition between the typical centimetre- and metre-regions of solar burst emission. By this, dm-bursts exhibit well known characteristics of both, the cm- and the m-bursts. On the other side there are also specific properties which are characteristical for bursts of the dm-region only. Such pecularities are the occurrence of intermediate drift bursts or typical fluctuations of radiation superposed on a continuum. The continua in the decimetre range sometimes appear simply as continuations of the centrimetric spectra, but there are also cases of special decimetric continua without resemblance to cm-emissions.

Two main phenomena are to be considered here. Firstly, the type IVdm-bursts challenge our attention, which sometimes are of a very high intensity ($5 \cdot 10^5$ s. u.) and of a comparatively small bandwidth ($\lesssim 500$ MHz). Concerning the occurrence of these characteristic type IVdm emissions the definition of the "dm-region" must be extended up to beyond 200 MHz. Nevertheless in many cases of type IV bursts a gap of decimetric components is observed resulting in a remarkable spectral minimum in this region.

Secondly, we have to deal with bursts of continuous radiation of lower intensity than the type IV bursts, quite similar to those of the common microwave bursts but extending into the dm-range. However, between the cm- and dm-regions a reversal of the sense of circular polarization is often observed. Furthermore there is a with wavelength increasing number of bursts of relatively small bandwidth. These bursts partly exhibit continuous and partly non-continuous radiation representing the transition from impulsive to drift bursts.

2.2.3. *Drift bursts*

Now some relations between continuous and non-continuous radio burst emissions are briefly considered so far as a joint origin is concerned. The bursts here under consideration are of the types I, II, III, and V of the spectral classification introduced by WILD (1950). These types, with partial exception of the first type, are characterized as drift bursts. The type V bursts which emit a pseudo-continuous radiation now are believed to be generated by a similar mechanism as the type III bursts.

Concerning the type II bursts (slow drift bursts) the relations to the occurrence of type IV bursts in the metre-region are of interest. On the other side connections to the type IVcm bursts are less clearly expressed. If the type II bursts are regarded as an indicator for the appearance of shock waves in the plasma of the solar corona, it can be concluded, that without doubt these shock waves often are associated with type IV bursts, but evidently they are neither a necessary nor a sufficient precondition for the occurrence of a continuous burst emission in the metre-region. In this way, shock waves as indicated by type II bursts have a greater importance for the study of the physics of the flare-burst phenomenon as a whole, than for an examination of only continuous burst radiation.

Bursts of the type III are much more frequent than type II bursts. As widely accepted, they can be used as an indicator for fast electron streams ($v \sim 0.5$ c). Because of the close association with flares and impulsive microwave bursts a joint primary origin of all these phenomena must be assumed. The height level in which the initial process leading to the flare-burst complex takes place appears to be variable within certain limits. However, it follows from statistical investigations that probably in most cases the place of the origin should be situated inside the coronal condensations. The energy of the electrons necessary for the interpretation of the type III bursts is of the order of 25 keV. As measured by satellite methods type III bursts are extending to at least 1 MHz, i.e. to a distance of about ten solar radii. The differing starting frequencies of the type III bursts can be explained as the effect of absorption in the deeper parts of the solar atmosphere depending on the energy of the electron stream (STURROCK, 1964b).

Type V bursts initially have been regarded as continuous emission sometimes following type III bursts at frequencies typically below 150 MHz (WILD et al., 1959). The perhaps only significant difference between the type V and the type III bursts consist in the prolonged duration of the emission of the type V bursts. In many cases type III and type V bursts are difficult to distinguish from each other. The first proposed idea that type V bursts could be explained by synchrotron emission was not tenable because of the required high particle energies ($\sim$2 MeV), which are not realized even for large pitch angles (STEWART, 1965). Statistical and theoretical studies have shown that a concept of continuous radiation is not necessary for an explanation of the type V burst emission. There are reasons for an interpretation of the type V bursts by plasma oscillations similar to that of type III bursts (DAENE and KRÜGER, 1966; WARWICK, 1967; ZHELEZNYAKOV and ZAJTSEV, 1968). The prolonged duration then may be explained by the special structures of the causing electron stream or by secondary processes taking place in the high corona.

2.2.4. *Type IV bursts*

In the general definition of type IV bursts as long period continuous radio burst emission accompanied by a major flare event (FOKKER, 1963; WILD, 1962) no restriction is made with respect to a special frequency range or other burst characteristics. As a consequence, depending on the spectral properties, the state of polarization, the directivity of the radiation, the movement of the burst source, etc., a number of subtypes of type IV bursts are to be distinguished. It turns out to be difficult to find the best way to establish such a subclassification so that the number of the proposals referring to this question not only reflects the complexity of the type IV burst phenomenon but also the number of prominent scientists working on this field. An attempt of a schematic representation of the hitherto known components of type IV bursts is demonstrated in fig. 6 in a form of a spectral diagram. For completeness also the other known spectral burst types are included in fig. 6.

2.2.4.1. Centimetre region

The effects of the type IV bursts in the cm-region are present in different ways. In general two basic structural forms of bursts are to distinguish. The first one is expressed by the occurrence of gradual bursts as an extension of the type IV burst continua in other spectral ranges, which a priori is not to be con-

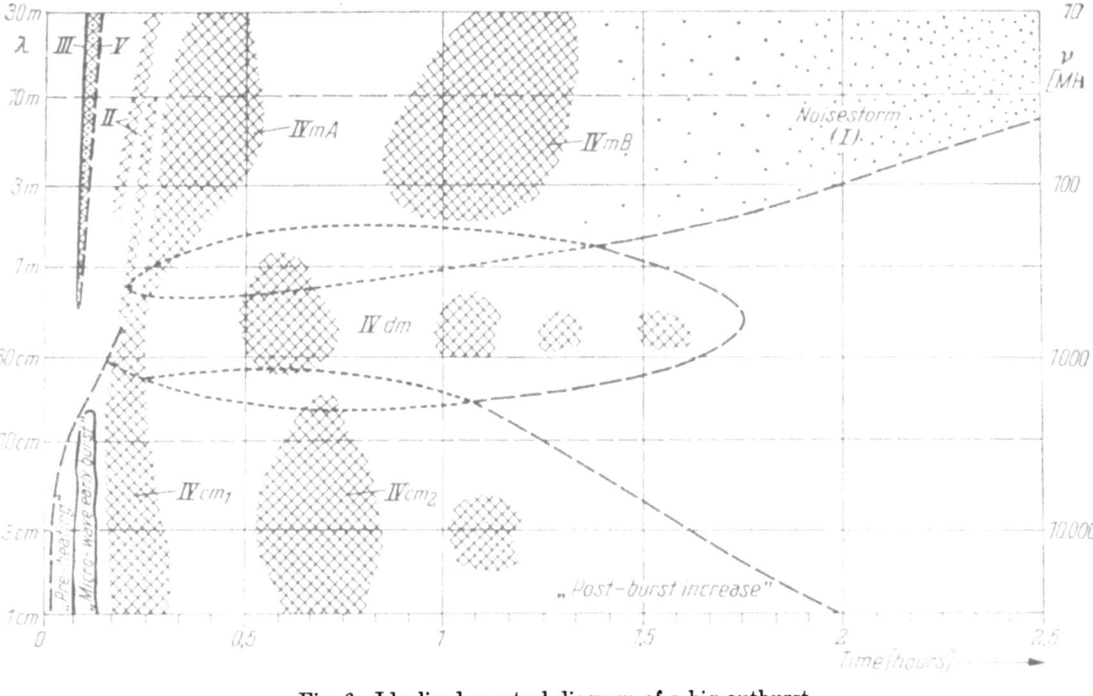

Fig. 6. Idealized spectral diagram of a big outburst

sidered as a proper type IV emission. In this connection it is not yet clear, whether gradual bursts associated with type IV bursts at longer wavelengths are physically distinct from actual gradual bursts without such an association. For gradual bursts the energy exchange is small, including the extremal case, when the spectrum is dropped below the limit of detectability, i.e. no centimetric component of a type IV burst is visible.

The other structural form is the actual type IVcm burst, which is often of a remarkable intensity. In the initial stage of the event the radiation sometimes steeply increases like characteristics of impulsive bursts. Usually several burst maxima are clearly expressed giving rise to an introduction of further subtypes (Tanaka and Kakinuma, 1962). In most cases the polarization corresponds to the extraordinary mode of the magneto-ionic theory if the leading spot hypothesis is accepted.

2.2.4.2. Decimetre range

As already mentioned in section 2.2.2 concerning the occurrence of type IVdm bursts the dm-range must be regarded as extending from above 200 to about 2000 MHz. This region is characterized by two remarkable features, namely the special polarization properties and the appearance of new structural details in the time profile of the burst radiation. The sense of circular polarization is often reversed to that of the type IVcm burst with some exceptions in the initial phase of the bursts. According to the magneto-ionic theory and assuming again the validity of the leading spot hypothesis, as usual the sense of the polarization of the type IVdm bursts corresponds to the ordinary mode, but also exceptions are possible. The relation between the sense of polarization[1]) and the magneto-ionic wave mode in dependence of the position on the Sun and the number of the solar cycle is given in table 2.

In the spectral diagram the type IVdm burst components typically exhibit narrow band patches of more or less intensity which are not necessarily directly related to type IV burst components in other spectral regions.

With respect to the above mentioned characteristics it must be concluded that the type IVdm bursts are no frequency- shifted analoga to the type IVcm bursts but a special phenomenon probably with own particle acceleration mechanism and a special generation mechanism of the radiation (Kai, 1965; Böhme, 1969).

2.2.4.3. The metre region

Due to the occurrence of moving and stationary phases the type IV burst phenomenon in the metre-region gets new aspects. The moving type IV burst components are indicated by interferometric detectable changes of the position of the burst source in the corona. These systematic changes of height and position of the centre of gravity of the observed burst radiation are not identical with

---

[1]) Definition cf. fig. 14

the rapid extension of the source region which practically always is present when radiation of a wide spectral region is emitted. These moving type IV burst components are the same phenomena as originally described and named "Type IV Burst" by BOISCHOT in 1957. Unfortunately up to now the moving type IV bursts are still not satisfactory investigated over the whole interesting spectral region. Especially the interrelation between special phases of the type IV burst complex and the expellation of plasma clouds and particles of different energies appears as a difficult question. From an investigation of moving type IV bursts in the spectral range 40–70 MHz it became evident that after an initial upward motion the burst source often moved downward and subsequently remained in a stable position in the corona (WEISS, 1963).

The moving type IV burst is predominantly polarized in the sense of the extraordinary mode of the magneto-ionic theory and has a relatively short duration. The directivity of the radiation is not very strong. With respect to these characteristics the moving type IV burst is identical with the type IVmA burst introduced by PICK-GUTMAN (1961) and used by various authors without concrete informations about the dynamics of the burst source.

The above mentioned characteristics of the type IVmA bursts are different from those of the stationary phase of type IV bursts denoted as type IVmB bursts. The smaller angular diameter of the source region and a sometimes observed frequency drift from higher to lower frequencies are somewhat puzzling features of the type IVmB bursts.

The close relation to the phenomenon of continuum bursts is of interest for an interpretation of the type IVmB bursts. Though distinct in the extremal phases of their development, no essential physical difference between the stationary type IV bursts and a commencing continuum storm hitherto could be stated. The latter sometimes develops into a regular type I noise storm. The sense of polarization is the same for both phenomena but the degree of polarization as usual subsequently rises with time.

2.3. Solar flares

It is a remarkable fact, that in principle all types of solar radio bursts are related to the flare phenomenon. Of course, there is no one-to-one correlation for all cases of flare and burst events, nevertheless the basic physical connection between the flares and the different types of burst radiation cannot be neglected. Merely the type I bursts and other forms of noise storm activity are initiated only sometimes by flare events.

Because of the high information content of the optical line emission and the better angular resolution, flare observations provide a wellcome completion to radio burst measurements. Many attempts have been made to explain the physical nature of the flare phenomenon on the basis of predominantly optical observations (SWEET, 1958; GOLD and HOYLE, 1960; SEVERNY, 1961). But it must be taken into mind, that the optical flare does not comprise the whole phenomenon. The $H\alpha$-radiation comes from regions where $T \lesssim 10^4$ °K, and therefore contains informations only about the outer zones of the whole flare disturbance. In contrast to this the radio bursts originate in regions of much higher temperatures covering different volumina. Therefore it can be expected that only a synthesis of optical and radio observations, as well as of measurements in other spectral regions and the particle emission can lead to a satisfactory explanation of the physical processes underlying the whole flare-burst phenomenon.

Because there are several monographs, proceedings of symposia, and review papers on the flare problem (e.g. SMITH and SMITH, 1963; HESS (ed.), 1964; DE JAGER and ŠVESTKA, 1969), in the following we intend to restrict ourselves to only some summarizing remarks which are of interest also from the point of view of the interpretation of solar continuum radio bursts.

2.3.1. *Flare characteristics*

At present it is still unknown, which flare characteristics or investigation will give a clue to a full physical understanding of the flare process. As a matter of fact, great difficulties of a physical interpretation arise from the suddenness ($\sim$100 sec) and the magnitude of the energy exchange ($\lesssim 10^{32} - 10^{33}$ erg) of the flare phenomenon. As it follows from optical and radio observations there must exist high gradients of temperature and electron density in that regions where the flare process takes place. This leads to the assumption of the essential influence of subtelescopic structures including turbulence (SUEMOTO and HIEI, 1959). But also macroscopic structures and motions are to be considered. It may be remembered, that e.g. loop prominences associated with flares are sometimes ascending as a whole and show downward motions of particle streams.

Many observations are in favour of a magnetic origin of the flare. Flares are observed preferably in locations where the gradient of the (photospheric) horizontal magnetic field $\frac{d H_L}{dx}$ is largely enough (e.g. $\nabla_x H_L \gtrsim 0.1$ gauss km^{-1} for proton flares) and the field H_L itself is small or zero ("neutral point"). This means that flares typically appear in regions between opposite magnetic polarities.

On the other hand, the transverse field component below flare regions show a characteristically inhomogeneous structure. Comparisons between the photospheric magnetic field maps before and after flare events show, that the flare process is connected with a simplification of the field structure and a decrease of the field gradients (SEVERNY, 1958).

During the flare process an extension of the flare area is observed also in directions perpendicular to the magnetic field. There are connections between flares and dark filaments which often are marking the position of neutral lines of the longitudinal magnetic field in complex spot groups (KIEPENHEUER, 1965).

Concerning the configuration of the magnetic field in the flare region hence a certain similarity to the conditions used in the theory of filaments by KIPPENHAHN and SCHLÜTER (1957) is present.

For a physical explanation of the flare mechanism it would be desirable to observe changes of the magnetic field immediately as a consequence of dissipative processes in the flare. But among other things, difficulties are met from the fact, that the optically derived magnetic field usually corresponds to deeper levels than that of the origin of the flare-burst event. One has also carefully to distinguish between changes of the magnetic fields due to dissipation processes and other non-dissipative field variations which normally are much greater than the small field changes expected from the flare effect. Nevertheless a systematic decrease of the gradients of the longitudinal magnetic field has actually been observed depending on the importance scale of the flare events (SEVERNY, 1969).

The production of flares appears as a common and quite regular feature in the development of complex solar active regions. The frequency of the occurrence is statistically determined by the following relation

$$N_F \sim A \cdot f\left(\frac{dN_F}{dt}\right), \qquad (2.3.1.\text{-}1)$$

where A — sunspot area and N_F — approximate number of flares per day (DE JAGER, 1965).

Observed connections between flare occurrences in distant active regions (WILD, 1969) and a preference of the production of type IV bursts for those cases in which more than one spot group is present on the Sun, lead to the conclusion that the single centres of activity are not isolated from each other but linked together by different kinds of interactions.

2.3.2. Flares with continuous emission in the optical spectral range

The analogon to the bursts of continuous emission can be considered in flares of white light (integral eruptions). Such flares are available only in a restricted number of cases. They hardly can be explained as a simple prolongation of the radio continuum, e.g. by assumption of synchrotron emission or bremsstrahlung as discussed by STEIN and NEY (1963). These mechanisms would require a great number of relativistic electrons which as a consequence of the inverse COMPTON effect must produce a much stronger X-ray emission than it is observed (KORCHAK, 1965). For this reason the actual emission mechanism of the white light flares remains yet unknown (ŠVESTKA, 1966).

2.3.3. Flares and radio emission

Because flares originate from brightenings of plage structures there are close relations to the plage phenomenon. Similar relations are present also between radio bursts (especially in the cm-region) and the S-component of solar

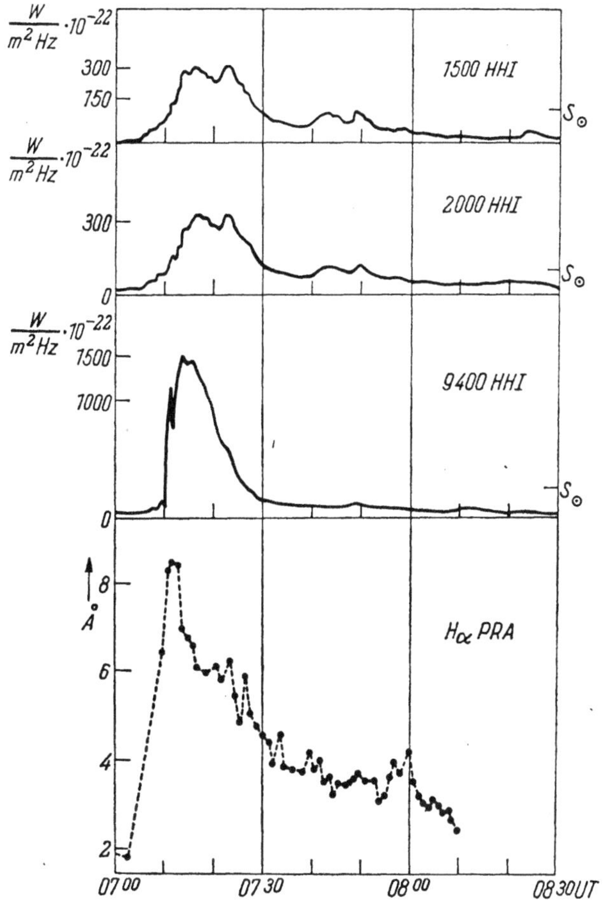

Fig. 7. Example for simultaneously observed microwave radio flux and $H\alpha$-line widths (1963, Sept. 26)

radio emission. There is in first order an approximately linear connection between the numbers of flare events per time interval and microwave bursts on the one hand and the plage areas or sunspot numbers on the other

$$N_F \sim N_B \; ; \qquad N_B \sim R_Z \, . \tag{2.3.3.-1}$$

Further similarities are evident between the phenomena of homologous flares and homologous bursts which may be considered as an indication of a retaining of special structures in active regions (Fokker, 1967).

Similarities can be observed also between the time profiles of the $H\alpha$-line width of the flares and single frequency records of microwave bursts (fig. 7). It is to be noted that to some extent the $H\alpha$-line width can be regarded as an indicator for the depth of the flare emission in the chromosphere. In the extreme cases of "moustaches" or "ELLERMAN bombs" the flare centre is situated in the lower chromosphere. However, if the $H\alpha$ emission is observed in the line centre as usual, the flare region is situated in heights greater than 5000 km above the photosphere and is extending into the lower corona.

2.4. Solar UV-, X-, and γ-radiations

Besides the emissions in the optical and radio spectral regions, solar centres of activity are also sources of an enhanced ultraviolet (in particular $L\alpha$) and X-radiation (especially in the region $\leqq 5 \cdots 60$ Å). There can be distinguished again a slowly varying component and burst radiations. Contrary to the behaviour of the S-component of the solar radio emission the slowly varying X-ray emission changes more rapidly and exhibit greater amplitudes amounting up to a factor of 80 and more in dependence of wavelength. In this way the background of the radiation of the quiet Sun radiation is practically neglectible. For this reason the slowly varying component of solar X-radiation is sometimes incorrectly called "quiet component" in contrast to the burst radiation.

The X-ray bursts are observed in a more or less close connection to flares and radio bursts (especially in the cm-region). With increasing equivalent temperatures of these X-ray bursts a shifting of the spectral maximum towards shorter wavelengths in the X-ray spectrum and a broadening of the emitted frequency range is observed.

The first indication of an increased sporadic X-ray emission associated with solar flares was indirectly concluded by observations of sudden ionospheric disturbances (SIDs) and geomagnetic solar flare effects (sfe's). As a striking feature the good correlation between the excessive sporadic solar X-ray emission and the occurrence of radio bursts in the short cm-region is well investigated. A criterion for the probability of an existence of a burst emission in both wavelength regions is provided by the exceeding of a limiting value of the integral

$$\int_{\text{Burst}} S_\nu \, dt$$

taken in the cm-region near about 3 cm wavelength (KRÜGER, 1963). In this criterion the fact is contained that either impulsive bursts or gradual bursts are possible in both wavelength regions.

Later, by the increasing possibilities of space technics more and more direct observations of solar X-rays became available and confirmed the close relationship between solar X-ray and radio emission. Reviews about the solar UV-, EUV-, and X-ray radiation have been given by TOUSEY (1963), FRIEDMAN (1963 and 1969), DE JAGER (1964), ALLEN (1965), MANDELSTAM (1965), TAKAKURA (1969) and others.

Concerning the γ-radiation up to present no radiation of a clear nuclear origin has been observed from the Sun. Regardless to the origin, short X-ray emission above 100 keV, i.e. below about 0.1 Å wavelength is sometimes called also γ-radiation. According to this definition solar γ-rays have been observed in rare cases up to about 0.5 MeV.

A summary of the main effects of solar flares in different wavelength and energy ranges is given in fig. 8.

2.5. Typical tendencies of the development of solar radio bursts and summary of the main aspects of phenomenological studies

The following main results of the phenomenological studies are to be emphazised:

a) There is a wide spectrum of solar flare effects being coupled together (fig. 8). This forces us to consider all these effects under a joint and uniform point of view. The connection between various different types of radio bursts is even more pronounced. This fact appears evident in particular for the phenomenon of the so-called compound bursts consisting of different types of radio bursts.

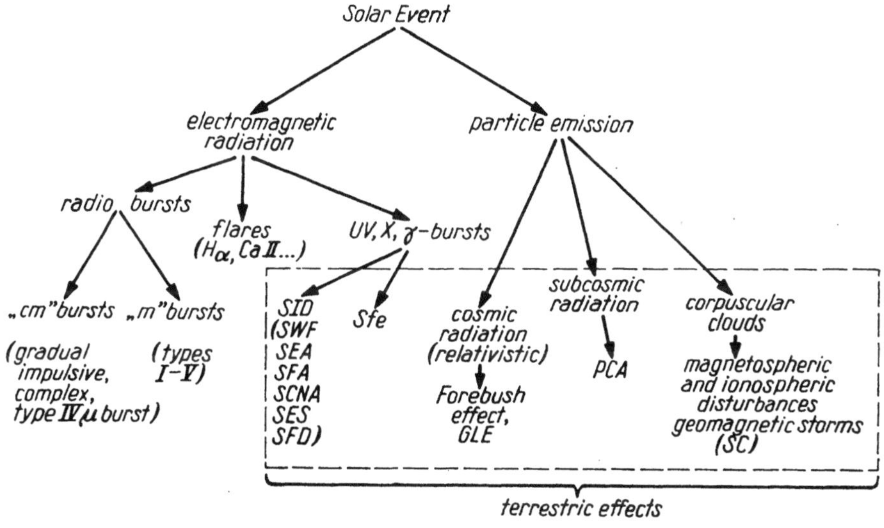

Fig. 8. Effects of solar flares

An example of such connections is established by a consideration of the morphological types of solar microwave bursts. Evidently a link between gradual bursts and superposed impulsive bursts can be assumed. On the other side, impulsive phases are also contained in complex radio bursts. Furthermore between complex bursts and type IV bursts phenomenologically only a quantitative, but no qualitative difference has been found. Thus, a circle consisting of three or four basic morphological types is closed, if it is noticed that type IV bursts in the upper cm- and lower dm-range are characteristically connected with gradual bursts at shorter wavelengths:

┌— type IV (complex) bursts ←┐
↓ |
gradual bursts ⟶ impulsive bursts

b) Besides the connections between fast (impulsive) and slow (gradual) burst phenomena also a relation between the general, slowly changing development of an active region and the generation of flares and bursts is present. Such relations are evident, when the development of the *S*-component and the occurrence of radio bursts is considered.

c) Summarizing the phenomenological features of all types of solar radio bursts and related phenomena, characteristical tendencies of the development of the burst events are detectable in spite of the great variety and complexity of individual cases.

Thus at first sight two main phases of the development of large radio bursts can be distinguished differing in the morphological characteristics and in the spectrum. The first phase is to be designated as the impulsive phase. To this phase there are belonging impulsive radio (microwave) and X-ray bursts as well as the flash phase of the optical flare. The second phase following the first one for large events comprises the different components of the type IV burst and other related second part events. In this scheme the type II burst has an intermediate character. Concerning its origin type II bursts are belonging to the first phase but with respect to the extension on the time scale type II bursts characterize the begin of the second phase.

An extension of the here considered characteristics of the flare-burst event as a whole and a more detailed classification of the different burst phases will be given following a theoretical discussion in chapter 4.

3. Theoretical foundations of the emission of radio waves in a plasma

3.1. Elementary processes of the generation of radiation

3.1.1. Single particle approximation

After the brief consideration of some of the main aspects of the burst phenomelogy in the previous chapter, now more theoretical discussions on the fundamentals of the generation of continuous radio emission are in the foreground of our considerations. At first a short summary of the basic emission mechanisms responsible for the emission of continuous radio waves will be given.

The primary emission processes to be considered here are due to interactions of particles and waves, i.e. we are dealing especially with the direct generation of radiation by electromagnetic transitions of accelerated particles (mainly leptons). Secondary emissions will be provided by different mechanisms of wave-wave interactions, in particular the transformation of longitudinal waves into transversal waves.

Concerning the primary emission processes, there are two large groups of emission mechanisms both belonging to the category of spontanous emission defined by Einstein so far as no amplification of radiation by negative absorption processes is considered.

The first group concerns free-free transitions of particles in an ionized medium, i.e. in particular bremsstrahlung by collisions of electrons in the Coulomb field of protons or positive ions. Free-bounded or bounded-bounded radiations have comparatively small effects in solar radio astronomy as it was shown by several authors (e.g. Hirth, 1963).

In contrast to the gyro-synchrotron process (see below) the bremsstrahlung mechanism is also possible in the absence of magnetic fields and typically the emission has no resonant character with regard to the selection of special frequencies. Nevertheless, its effiency as expressed by the optical depth can be largely enhanced near the zero regions of the complex refractive index ("absorption edges" of the dispersion relation).

The second, and from our point of view most important, group of emission mechanisms is represented by the gyro-synchrotron (cyclotron) radiation, sometimes also called magneto-bremsstrahlung, of electrons spiraling in a magnetic field. Its realization therefore is intimately related to the existence of magnetic fields and accelerated particles which both without doubt are present in solar active regions. As a characteristic feature the gyro-synchrotron process is most effective at special poles of the dispersion relation determined by the gyro frequency and its harmonics expressing the resonant character of the emission.

A limiting case of the synchrotron emission for vanishing external magnetic fields is established by the Čerenkov emission. But also for magnetic fields unequal zero there is a possibility of Čerenkov emission under certain circumstances, especially if the particle speed is directed parallel to the magnetic field. However, for applications in solar radio astronomy the direct emission of electromagnetic waves by the Čerenkov effect is less important than the the synchrotron emission and the the Čerenkov emission of plasma waves occurring in a hot plasma.

The most essential physical parameters for the action of the above mentioned emission mechanisms are the particle energy or velocity $v = \beta c$ and the strength of the external magnetic fields H. A rough classification scheme of the emission mechanisms according to these parameters is given in table 4.

Tab. 4. Elementary processes of emission

	$\beta \ll 1$ ("quasi-thermal")	$1 - \beta \ll 1$ ("non-thermal")
$H \sim 0$	Bremsstrahlung	Čerenkov
$H < 0$	Gyro-synchrotron emmission	radiation

Besides the direct emission of radio waves by particle-wave interactions the emission of the already mentioned plasma waves is of some interest. Because plasma waves are bound on a medium they cannot leave the solar atmosphere. For a detection on the earth the transformation of the plasma waves into electromagnetic waves by mechanisms of wave-wave interactions appears to be necessary.

It should be further mentioned, that in general the emission mechanisms listed in table 4 in solar burst phenomena are not realized isolately. In many cases a superposition of several components of radiation must be expected, the seperation of which mets to certain difficulties.

3.1.2. *Radiations from definite volumes*

After the clarification of the elementary emission processes of single particles next the emission of an ensemble of particles from definite volumina is to be calculated. The solution of this task includes the consideration of

a) the particle-particle interaction of the radiating particles,

b) the interaction of particles and waves with the ambient medium, and

c) the integrations over the energy distribution of the particles and the radiating volume.

In the simplest case of a first approximation the points a) and b) can be neglected and the velocity distribution of the emitting particles is identical with a MAXWELL distribution

$$f(v_x, v_y, v_z)\, dv_x\, dv_y\, dv_z = N \left(\frac{m}{2\pi k T}\right)^{3/2} \exp\left[-\frac{\frac{m}{2}(v_x^2 + v_y^2 + v_z^2)}{k T}\right] dv_x\, dv_y\, dv_z\,. \tag{3.1.2.-1}$$

This case refers to isotropy and thermal equilibrium. For applications to solar radio bursts, however, in most cases this approximation is not sufficient and deviations from the thermal equilibrium must be taken into account. Such deviations are present regarding the mean energy or distribution of the absolute values of the velocities $f(v)$, where $v = (v_x + v_y + v_z)^{1/2}$, but also with respect to anisotropies of the distribution of the velocities v_x, v_y, and v_z.

Examples for distributions describing deviations from thermal equilibrium conditions are given by e.g. MAXWELL distributions superposed with peaks of superthermal velocities (e.g. runaway electrons) or also by exponential and power-law energy distribution functions. The approximations of exponential and power law functions are valid only in special more or less extended regions of the energy spectrum.

It is to be noticed that the energy distribution function represents a solution of an equation of balance in which acceleration and deceleration processes are to be taken into account. In general, the energy or velocity distribution is strongly time dependent and variable in space depending on different initial and boundary conditions. This results in a complexity of possible cases which is reflected by the multitude of realizations of individual burst events. Therefore a restriction is necessary to a consideration of the main phenomena and basic physical processes referring to average conditions.

3.2. Bremsstrahlung without external magnetic fields

3.2.1. The equation of radiation transfer

The theory of radiation transfer (CHANDRASEKHAR, 1950; KOURGANOFF and BUSBRIGDE 1952) was developed within the framework of geometrical optics referring to regions which are large in comparison to the wavelength. The radiation exchange in a volumen element dV is given by the equation of transfer wich for an inhomogeneous medium with a variable refractive index n_j may be written as

$$n_j^2 \frac{d}{ds}\left(\frac{I_{\nu,j}}{n_j^2}\right) = -\varkappa_{\nu,j}\, I_{\nu,j} + \varepsilon_{\nu,j} \tag{3.2.1.-1}$$

where $I_{\nu,j}$ — intensity of radiation for an observing frequency ν in the direction of s, ds — element of the ray path, $\varkappa_{\nu,j}$ — absorption coefficient, $\varepsilon_{\nu,j}$ — emission coefficient. The index j refers to the possibility of different wave modes. Equation (3.2.1.1) can be regarded as a one-dimensional approximation of a general tensor equation of radiative transfer discussed later.

For a non-isotropic medium (e.g. under the presence of an external magnetic field) the term n_j^2 is to be replaced by the expression n_j^2 (cos ϑ_j), where

$$\vartheta_j = \sphericalangle \left(\boldsymbol{k}_j, \frac{d\omega}{d\boldsymbol{k}_j} = \boldsymbol{v}_{gr} \right) \tag{3.2.1.-2}$$

denotes the angle between the direction of the POYNTING vector s and the wave normal $\boldsymbol{k}_j$ (cf. fig. 9).

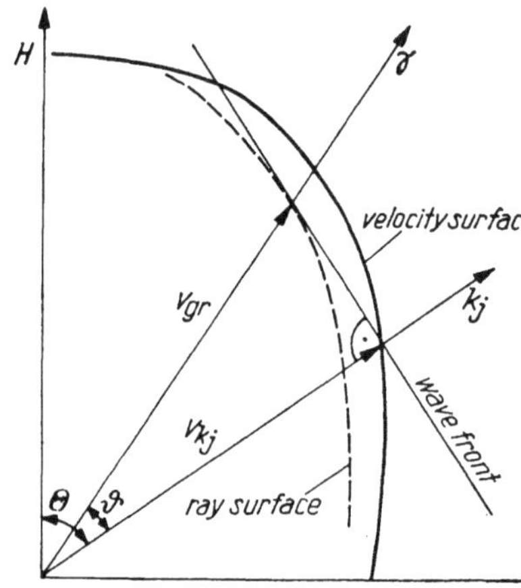

Fig. 9. On the geometry of wave front, phase velocity surface, and ray (i.e. group velocity) surface in the anisotropic plasma

With introduction of the optical depth

$$\tau_{\nu,j} = \int_0^\infty \varkappa_{\nu,j}\, ds \tag{3.2.1.-3}$$

the equation of radiation transfer is written

$$\sec \vartheta_j \frac{d}{d\tau_{\nu,j}} \left(\frac{I_{\nu,j} \cos \vartheta_j}{n_j^2} \right) = - \frac{I_{\nu,j}}{n_j^2} + \frac{\varepsilon_{\nu,j}}{\varkappa_{\nu,j}} \frac{1}{n_j^2} \,. \tag{3.2.1.-4}$$

The second term of the right side of (3.2.1.-4) is called ergiebigkeit or source function which determines the intensity of the radiation.

The above given expressions for the equation of transfer are also valid for cases without radiative equilibrium (the condition of radiative equilibrium is only expressed by the equation of continuity).

Applications of the here given approximations of the equation of radiation transfer are most convenient if polarization effects are not to be taken into consideration. For the transfer of polarized radiation the tensor theory of radiative transfer is more likely to be used (cf. section 3.4.3.).

3.2.2. *The absorption coefficient*

In the equations of the previous section the emission and absorption coefficients are contained as essential quantities. Now we are dealing with some methods of the calculation of these coefficients, especially the absorption coefficient in the case of bremsstrahlung.

Basing on the elementary emission process (electron — proton collision) one method of calculation of $\varkappa_{\nu,j}$ starts with the evaluation of $\varepsilon_{\nu,j}$. This method implies the calculation of the emission of a single electron radiating in the COULOMB field of a proton (or positive ion) with the subsequent integration over the collision parameter and the velocity distribution of the ecletrons. Then the absorption coefficient is derived from KIRCHHOFF's law assuming thermal equilibrium.

Calculations by this method can be found by KRAMERS (1923), UNSÖLD (1955), OSTER (1961) and others.

For burst physics the application of the values of the absorption coefficient derived in this manner is limited because of the restriction to thermal equilibrium and due to the fact, that the important relation to the (real) refractive index is primarily not contained in this theory ($n_j = 1$). Its consideration follows from a subsequent artificial study. Another difficulty arises from the question up to which values of the collision parameter the integration must be carried out. Thus different authors obtain slightly different values for the absorption coefficient mainly expressed by the special form of the COULOMB-logarithm. Special results are

$$\varkappa_\nu^{\text{klass}} = N_i\, N_e \frac{8\, e^6\, Z_k^2}{3\sqrt{2\pi}\, c\,(m\, kT)^{3/2}\, \nu^2} \ln\left\{\frac{kT}{1.44\, Z_k\, e^2\, N_e^{1/3}}\right\}$$

$$= 1.318 \cdot 10^{-2} \frac{N_e^2}{T^{3/2}\, \nu^2} \ln\left(367 \cdot T\, N_e^{-1/3}\right) \qquad (3.2.2.\text{-}1)$$

(UNSÖLD, 1955)

$$\varkappa_\nu^{\text{klass}} = 1.32 \cdot 10^{-2} \frac{N_e^2}{T^{3/2}\, \nu^2} \ln\left(4.27 \cdot 10^7\, T^{3/2}\, \nu^{-1}\right) \qquad (3.2.2.\text{-}2)$$

(OSTER, 1961)

Instead of the classical computation of the emission coefficient the calculation is also possible on the basis of quantum physics. In this case certain corrections of the classical calculation are obtained:

$$\varkappa_\nu^{\text{quant}} = 1.32 \cdot 10^{-2} \frac{N_e^2}{T^{3/2}\, \nu^2} \ln\left(4.68 \cdot 10^{10}\, T\, \nu^{-1}\right) \qquad (3.2.2.\text{-}3)$$

(OSTER, 1961)

A completely other approach to the calculation of $\varkappa_\nu$ starts from the LORENTZ theory of electrons, or, its extension, the magneto-ionic theory. In the framework of a hydrodynamic consideration the differential equation of the movement of an "average" electron is employed. From this a refraction index is obtained which becomes complex if collisions are taken into account. According to the dispersion theory the imaginary part of the refraction index multiplied by the factor ω/c is identical with the absorption coefficient.

This method of calculation has been used for the estimation of bremsstrahlung in the presence of external magnetic fields (cf. also 3.3.4). A certain methodological difficulty, however, arises from the introduction of the collision frequency which in the theory of a cold plasma originally is not present and comes formally from the assumption of a "term of friction" in the basic equation of the electron motion.

By this treatment the absorption coefficient gets the following form

$$\varkappa_\nu^{\text{magn.-ion}} = \frac{\omega}{c}\left\{\frac{[(1 - X + Z^2)^2 - X^2 Z^2]^{1/2} - [(1 - X) + Z^2]}{2\,(1 + Z^2)}\right\}^{1/2}$$
$$= \frac{\omega}{c}\left\{-\frac{1}{2}\left(1 - \frac{\omega_p^2}{\omega^2 + \nu_{\text{coll}}^2}\right) + \frac{1}{2}\left[1 - \frac{\omega_p^2}{\omega^2 + \nu_{\text{coll}}^2}\right]^2 + \right.$$
$$\left. + \left(\frac{\omega_p^2}{\omega^2 + \nu_{\text{coll}}^2} \cdot \frac{\nu_{\text{coll}}}{\omega}\right)^2\right]^{1/2}\right\}^{1/2}, \tag{3.2.2.-4}$$

where X and Z are the parameters used in the magneto-ionic theory

$$X = \left(\frac{\nu_p}{\nu}\right)^2, \quad Z = \frac{\nu_{\text{coll}}}{2\,\pi\,\nu}$$

ν_p — plasma frequency, ν_{coll} — collision frequency. For $X \ll 1$ and $Z \ll 1$ the well known expression follows

$$\varkappa_\nu = \frac{\omega}{c}\,\frac{X\,Z}{2\sqrt{1 - X}} = \frac{1 - n^2}{c\,n}\,\nu_{\text{coll}}\,. \tag{3.2.2.-5}$$

As mentioned above, all the expressions given here refer to the approximation of a cold plasma, i.e. the thermal motion of the particles in the medium is not taken into account.

At last we refer to the general quantum-physical calculation of the absorption coefficient using the EINSTEIN coefficients $A_{E\to E-h\nu}$, $B_{E\to E-h\nu}$, and $B_{E-h\nu\to E}$ for spontanous emission, induced emission, and absorption, respectively. This calculation is not limited to the special case of bremsstrahlung but depends on the condition of detailed balance.

There results the relation

$$\varkappa_\nu = -\frac{e^2\,N_e}{n^2\,\nu^2}\int\limits_0^\infty \frac{dF(E)}{dE}\cdot Q_\nu(E)\cdot g(E)\,dE\,, \tag{3.2.2.-6}$$

where referring to radio waves $h\,\nu \ll E$ has been assumed. $F(E)$ denotes the distribution function of electron energies, Q_ν is the emissivity of an electron of the energy E per unit time and frequency interval in one polarization per unit solid angle into a given direction. $g(E)$ are statistical weights of the energy levels E.

Analogously the emission coefficient is given by

$$\varepsilon_\nu = N_e \int_0^\infty F(E) \cdot Q_\nu(E) \cdot g(E)\, dE \, . \tag{3.2.2.-7}$$

According to the correspondence principle of BOHR the classical calculation again can be obtained as limiting case of the quantum theoretical calculations (ELWERT, 1948).

3.2.3. Stationary thermal burst models

The simplest case applies to radiation fields which are in equilibrium with the ambient medium. Then the emission and absorption processes are connected according to KIRCHHOFF's law

$$\frac{\varepsilon_\nu}{\varkappa_\nu} = I_\nu(\nu, T) \tag{3.2.3.-1}$$

and the source function (ergiebigkeit) is identical with the KIRCHHOFF-PLANCK function

$$B_\nu(T) = \frac{2\,k}{c^2}\,\nu^2\, T \, . \tag{3.2.3.-2}$$

Because in real solar conditions the thermal equilibrium is not exactly realized the conception of the local thermodynamical equilibrium was introduced (SCHWARZSCHILD, 1914 and MILNE, 1921). Instead of the isothermal temperature T, a local temperature $T(x, y, z)$ is defined for which the formalism of the thermal radiation is employed. Here again the assumption of a radiative equilibrium is not necessarily needed. Besides radiation, also other processes of energy transfer, e.g. convection and heat conduction may contribute to a maintenance of the temperature distribution.

Often also more or less serious deviations from the local thermodynamical equilibrium are treated by the same method. In these cases the radiation is called "quasi-thermal".

Thermal or quasi-thermal bremsstrahlung have been proposed for the interpretation of certain components of the burst radiation in the microwave region (HACHENBERG and WALLIS, 1961; KRÜGER, 1961, GELFREIKH, 1962a, b; HIRTH, 1963, KAWABATA, 1963). Though solar radio bursts must be regarded as strongly non-thermal phenomena, at least partly for one component of the burst emission, namely the gradual bursts an effect of thermal radiation a priori cannot be excluded.

In the following the fundamentals of a thermal interpretation are shortly sketched and critically analyzed. With regard to (3.2.3.-2) the equation of radiation transfer gets the simple form

$$\frac{d}{d\tau_\nu}\left(\frac{I_\nu \cos\vartheta}{n^2}\right) = -\frac{I_\nu \cos\vartheta}{n^2} + \frac{2\,k\,\nu^2\, T}{c^2} \, . \tag{3.2.3.-3}$$

The solution of this inhomogeneous differential equation of first order is given by

$$\frac{I_\nu \cos\vartheta}{n^2} = e^{-\tau_\nu}\left[\frac{I_\nu \cos\vartheta}{n^2}\right]_0 + e^{-\tau_\nu}\int\limits_0^{\tau_\nu} \frac{2\,k\,\nu^2\,T}{c^2}\,e^{-\tau_\nu}\,d\tau_\nu\,. \qquad (3.2.3.\text{-}4)$$

From this a number of wellknown special cases can be derived. For the radiation coming from an emitting layer of temperature T and optical depth $(\tau_{\nu,0} - \tau_\nu)$ into a vacuum it follows

$$I_\nu = I_{\nu,0}\,\frac{n_0^2}{\cos\vartheta_0}\,e^{\tau_{\nu,0}-\tau_\nu} + 2\,\frac{k\,\nu^2}{c^2}\,T\,(1 - e^{\tau_{\nu,0}-\tau_\nu})\,. \qquad (3.2.3.\text{-}5)$$

Hence one obtains with $n = \text{const}$, $\cos\vartheta = \text{const}$, and $T \neq \text{const}$

$$I_\nu = e^{-\tau_\nu}\,I_{\nu,0} + e^{-\tau_\nu}\,\frac{n^2}{\cos\vartheta}\int\limits_0^{\tau_\nu}\frac{2\,k\,\nu^2}{c^2}\,e^{\tau_\nu}\,d\tau \qquad (3.2.3.\text{-}6)$$

and, if $n = 1$, $\cos\vartheta = 1$, and $T = \text{const}$, it follows

$$T_\nu = I_{\nu,0}\,e^{-\tau_\nu} + B_\nu(\tau)\,(1 - e^{-\tau_\nu})\,. \qquad (3.2.3.\text{-}7)$$

This case was applied for the calculation of the thermal radio emission of coronal condensations (S-component) by WALDMEIER and MÜLLER (1950), IKHSANOVA (1960), and others. According to (3.2.3.-6) and (3.2.3.-7) the brightness temperature can be written as

$$T_s = e^{-\tau_\nu}\,T_{s,0} + e^{-\tau_\nu}\int\limits_0^{\tau_\nu} T\cdot e^{\tau_\nu}\,d\tau_\nu \qquad (3.2.3.\text{-}8)$$

or

$$T_s = e^{-\tau_\nu}\,T_{s,0} + T\,(1 - e^{-\tau_\nu})\,. \qquad (3.2.3.\text{-}9)$$

From the equations (3.2.3.-7) and (3.2.3.-9) the well known relations for the limiting cases of very large and very small optical depths follow

$$\left.\begin{aligned} \tau_\nu \gg 1 \searrow I_\nu &= B_\nu(T)\,, \quad T_s = T\,, \\ \tau_\nu \ll 1 \searrow I_\nu &= B_\nu(T) + I_{\nu,0}\,(1 - \tau_\nu)\,, \end{aligned}\right\} \qquad (3.2.3.\text{-}10)$$

$$T_\nu = T\cdot\tau_\nu + T_{s,0}\,(1 - \tau_s)\,. \qquad (3.2.3.\text{-}11)$$

By this e.g. it is evident that only under very special conditions (homogeneity, isotropy and isothermal medium) a spectral curve parallel to the frequency axis is resulting, in all other cases the spectrum is inclined (cf. below).

Basing on the knowledge of the emission process, a fundamental problem of solar radio physics consists in the derivation of the plasma parameters, e.g. temperature and density, in the source region from the observed radiation characteristics, being represented by the spectrum of the flux density

$$S_\nu = \int\limits_{\text{source}} I_\nu\,d\Omega \qquad (3.2.3.\text{-}12)$$

(Ω — solid angle) by means of theoretical models for the spatial (and temporal) distribution of I_ν or τ_ν.

Besides the spectral characteristics, the dimensions of the source region is another important quantity, which in many cases is not well observed. But even, if the projected seize of the source region and the distribution of the intensity in it at any observing frequency are measured, it must be stated that the above formulated problem in general cannot be definitely solved, because the third dimension, i.e. the dependence in height of the radiation remains principally unknown. Therefore, additional data about the centre-limb variation of the radiation and results of observations in other spectral regions must be used and special model assumptions are to be introduced.

At first a rough summary of the conclusions following from a thermal interpretation may be given. In fig. 10 the magnitudes of the effective temperatures T_e for an observing frequency of 10 GHz (3 cm wavelength) are plotted in dependence of the seize F of the emitting region (expressed in millionths of the solar disk) for black body radiation, which is based on the relation

$$S_\nu = \frac{2k}{c^2} F\, \Omega_\odot\, T_e \nu^2 = 0.188 \frac{F \cdot T_e}{\lambda^2_{[\mathrm{cm}]}} \text{ [Solar units]} = \left[\frac{W}{m^2 H_z}\right] \cdot 10^{-22}\,. \tag{3.2.3.-13}$$

For the case of very small optical depths the ordinate scale must be multiplied by the factor τ_ν according to equation (3.2.3.-11). For comparison, the spectrum of the quiet Sun is included in fig. 10.

As it follows from the solution of the transfer equation and the expression of the absorption coefficient for thermal bremsstrahlung, without presence of

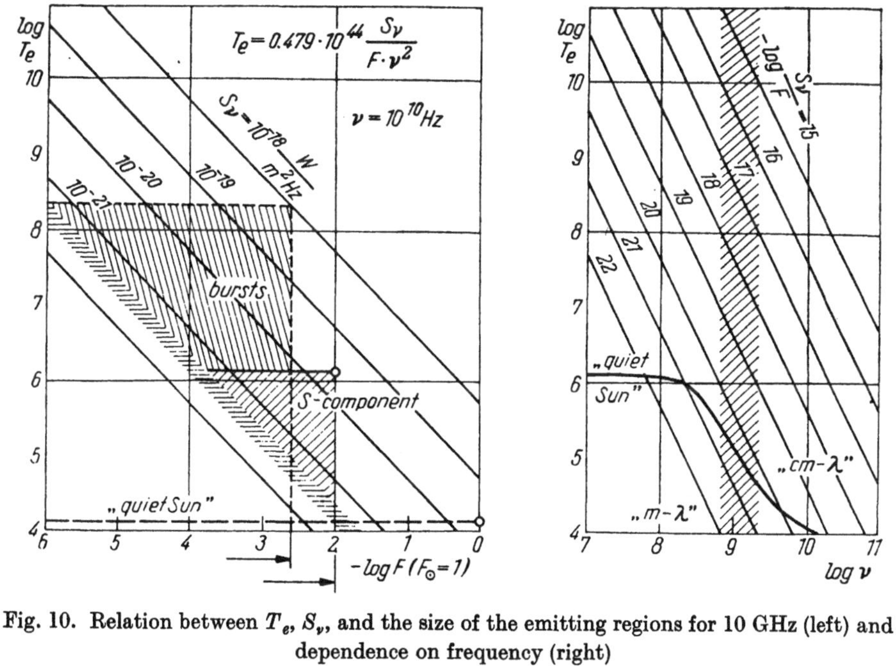

Fig. 10. Relation between T_e, S_ν, and the size of the emitting regions for 10 GHz (left) and dependence on frequency (right)

magnetic fields only spectra of I_ν increasing monotonously towards higher frequencies are possible (for an exception see the next section).

As mentioned above, in the limiting case of very small optical thickness under the assumption valid for equation (3.2.3.-7), I_ν becomes independent of the frequency. In this case the spectrum is basically determined by the electron density, which can be estimated by the following relation

$$N_e = \left\{ \frac{0.363 \cdot 10^{46}\, T_e^{1/2}\, S_\nu}{F \cdot \ln\left(4.68 \cdot 10^{10} \frac{T}{\nu}\right) L} \right\}^{1/2} [\mathrm{cm}^{-3}] \qquad (3.2.3.\text{-}14)$$

where L is a characteristical height extension of the emitting volume.

In the other limiting case of large optical thickness, the slope of the spectrum is proportional to ν^2. Hence an estimation of the effective temperature becomes possible:

$$T_e = \frac{0.479 \cdot 10^4\, S_\nu}{F \cdot \nu^2} [^\circ\mathrm{K}]\,. \qquad (3.2.3.\text{-}15)$$

Between both limiting cases of great and small optical depth a great number of intermediate cases exist influencing the slope of the spectrum of radiation intensity. A flattening of the slope of the spectrum of intensity of flux density must be further expected, if the medium is not isothermal and the temperature rises with increasing height. The opposite case of temperature gradients exhibiting growing temperatures towards deeper levels leads to a steepening of the spectrum.

The shape of the spectrum is furthermore influenced by external magnetic fields (cf. 3.3.4). In all these cases at each frequency the intensity of the black body radiation for a corresponding temperature cannot be exceeded.

3.2.4. Influences of the plasma frequency

According to equation (3.2.2.-4) near the plasma frequency (which was not considered in the foregoing sections) special conditions are present. Besides the cut-offs for the propagations of lower frequencies (cf. 3.3.1), in this region even without the presence of external magnetic fields a maximum in the spectrum of intensity or flux density can occur. As it was shown by HIRTH (1963) and HIRTH and WALLIS (1966), for the existence of such a maximum the condition

$$G = \frac{\nu_{\mathrm{coll}}\, L}{c} < 1 \qquad (3.2.4.\text{-}1)$$

must be fulfilled, where L is the height of the emitting layer (c — velocity of light).

The reached relative value of the intensity of the maximum is then

$$\zeta' = \frac{I_{\nu_{p,\max}} - I_{\nu_0}}{I_{\nu_0}} \approx G^{-1} - 1\,, \qquad (3.2.4.\text{-}2)$$

where

$I_{\nu_{p,\max}}$ is the intensity at the frequency of the spectral maximum and
I_{ν_0} is the intensity at frequencies where $\tau_\nu \ll 1$.

Assuming special models of the emitting region a similar expression for the flux density can be derived. For instance with a lens-like model of the distribution of the electron density one obtains the maximum of the observed flux density as

$$S_{\nu,\max} = \left(\frac{r_0}{R_{\odot \delta}}\right)^2 \frac{2\pi \nu_{\max}^2 kT}{c^2} \qquad (3.2.4.\text{-}3)$$

and analogously to (3.2.4.-2)

$$\eta' = \frac{3c}{\nu_{\text{coll}} L} \qquad (3.2.4.\text{-}4)$$

with r_0 — the radius of the emitting region and L — maximal height of the emitting region.

If one attempts to interprete a microwave emission by this effect for which e.g. at 10 cm wavelength $S_{\nu,\max}$ amounts to, say, 30 solar flux units, there would follow electron densities N_e of the order of 10^{11} cm^{-3} and heights of the emitting layer L of at best 10^3 km. The latter is a very small value for L and poses the question of subtelescopic fine structures in the source regions of radio bursts. Because of the lack of suitable measurements up to present it remains open, whether the above mentioned effect actually takes place in radio bursts or not. In this way, the remarks given here can only draw the attention to some effects theoretically expected near the plasma frequency needing the confirmation by observations as well as further improvements by calculations on the basis of kinetic theory.

3.2.5. Preliminary interpretation of the morphological types of solar microwave bursts

In the previous sections the special shape of the time profile of the single burst events has not been taken into account. Now the consideration of the instationary behaviour of the burst phenomena will be considered which leads to some further problems concerned with special forms of energy supply, the different loss processes, motions of matter etc. Basing on these processes the electron density and electron energy appear as essential quantities determining the emitted radiation in each time interval. The particle energy is described by suitable distribution functions (cf. 3.1.2) the maximum of these distributions mainly determines the value of the equivalent temperature, so far as no great deviations from a Maxwellian distribution are concerned.

Furthermore the life time of the burst radiation is of special interest. In fig. 11, according to TAKAKURA (1960), the life time of the radiation is shown in dependence of the electron density and energy due to eneryg losses by collisions and synchrotron emission (see section 3.5.1).

Now we try to interprete preliminarily and very tentatively the basic morphological types of centimetric bursts in dependence of the magnitudes of the particle density (and emission volume) and the energy (velocity distribution). With respect to these parameters roughly four cases can be distinguished which are plotted in table 5.

Small particle numbers and energies are expected to be associated with small microwave bursts of short duration. Small energy and larger particle numbers

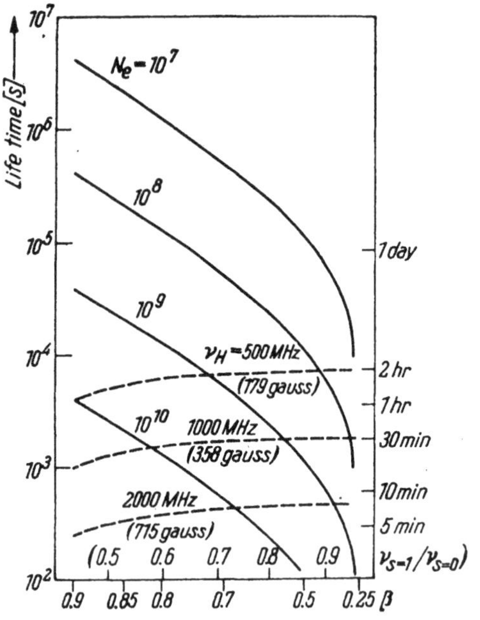

Fig. 11. Life times of the radio emission due to radiation losses of cyclotron radiation (solid curves) and collision losses with thermal electrons (dashed curves) after TAKAKURA (1960)

Tab. 5. On the interpretation of the main morphological types of microwave bursts

Electron density and emission volume	particle energy	
	small	large
small	small bursts	impulsive bursts
large	gradual bursts	transition to type IV bursts

may cause bursts of the type gradual rise and fall. On the other side, high energies and small numbers of emitting particles would lead to impulsive bursts, whereas both, high energies and large particle numbers are causing larger, mostly complex bursts which may be referred to as type IV bursts.

It must be noted, that the scheme above given is only a very rough attempt of obtaining a zero order approximation of the reality. Especially a more detailed consideration of the forms of the acceleration processes, the action of plasma streams and shock waves, influences of magnetic fields on the radiation and trapping of particles connected with complicated temporal and spatial structures appears necessary in order to improve our interpretation of the burst morphology. Nevertheless, besides the general idea, table 5 provides also an impression about the applicability and the limitation of a thermal interpretation which is expected to be restricted at best only to the first column of table 5, i.e. to small microwave bursts and gradual bursts.

3.3. Bremsstrahlung in the presence of external magnetic fields

3.3.1. Propagation of electromagnetic waves in a plasma

Before considering the modifications of the emission by external magnetic fields, a short compilation of some fundamentals of the propagation of radio waves in ionized media will be given.

On the basis of MAXWELL's field equations combined with the BOLTZMANN or VLASOV equation (kinetic theory) or the equations for charge and momentum conversion (transport theory) dispersion relations are derived describing the propagation characteristics of waves in a plasma. Usually a restriction is made to waves of small amplitudes avoiding complicated non-linear mathematics, as it would be needed for a treatment of waves of arbitrary amplitudes, e.g. shock waves.

The kinetic theory describes the general case of waves in a plasma taking into account the motions of ions and electrons. If, on the other side, instead of the BOLTZMANN equation, the equation of single particle motion is used, the periodic solutions of the resulting system of equations are related to (radio) waves in a cold plasma (magneto-ionic theory). But if the (linearized) hydrodynamic equations of motion (e.g. for a two-component plasma) are used, dispersion relations for the magneto-hydrodynamic waves are obtained in which the electromagnetic waves are contained as limiting case.

In any case, the dispersion relations are obtained as condition for the solubility of the resulting equation system, which can be written in form of a tensor eigen value equation. The thus obtained general dispersion relation is of the fourth degree in n^2 (refractive index). The solution refer to four wave modes, which may be called the ordinary and extraordinary magneto-hydrodynamic waves, the magneto-acoustic and the magnetic plasma waves (PIDDINGTON, 1955).

If the ion motion is neglected, the magneto-acoustic waves vanish and the "ordinary" and "extraordinary" electromagnetic waves are obtained as limiting case of the magneto-hydrodynamic waves. The dispersion relation then can be written in the following form

$$\beta_{\text{th}}^2 (1 - Y \cos^2 \theta)\, n^6 - [1 - X - Y + X\, Y \cos^2 \theta + \\ + 2\, \beta_{\text{th}}^2 (1 - X - Y \cos^2 \theta)]\, n^4 + [2\,(1 - X)^2 - Y\,(2 - X - X \cos^2 \theta) + \\ + \beta_{\text{th}}^2 (1 - 2\, X + X^2 - Y \cos^2 \theta)]\, n^2 + (1 - X)\,[Y - (1 - Y)^2] = 0 \tag{3.3.1.-1}$$

where

$$X = \left(\frac{\omega_p}{\omega}\right)^2, \quad Y = \frac{\omega_H}{\omega}, \quad \theta = \sphericalangle\, (\boldsymbol{k}, \boldsymbol{H}) \tag{3.3.1.-2}$$

and

$$\beta_{\text{th}} = \left(\frac{3\, k\, T}{m_e\, c^2}\right)^{1/2} \tag{3.3.1.-3}$$

is a dimensionless quantity of the magnitude of the quotient of the mean thermal electron velocity and the velocity of light.

Equation (3.3.1.-1) is of third degree in n^2. It describes both transversal wave modes of the electromagnetic waves and the plasma waves as a longitudinal wave type. If the thermal motion of the electrons is neglected ($\beta_{\text{th}}^2 = 0$), also the plasma waves vanish and the fundamental equation of the magneto-ionic theory is obtained

$$n_j^2 = 1 - \frac{2\, X\,(1 - X)}{2\,(1 - X) - Y_T^2 \mp [Y_T^4 + 4\,(1 - X)^2\, Y_L^2]^{1/2}} \tag{3.3.1.-4}$$

where $Y_L = Y \cos \theta$ and $Y_T = Y \sin \theta$ denote the longitudinal and transversal components of Y with regard to the magnetic field H, respectively. The index 1 (upper sign) refers to the extraordinary wave and the index 2 to the ordinary wave (lower sign), respectively.

However, it must be mentioned, that the denotation of the wave modes is not uniform in the literature. In ionospheric research and radio astronomy usually the notations ordinary and extraordinary wave have been used for the whole branchs of the dispersion relation. Contrary to this, in certain monographs on plasma physics (cf. STIX, 1962; ALLIS, BUCHSBAUM, BERS, 1963) these notions are only applied for the case $\theta = \frac{\pi}{2}$. For arbitrary angles θ the notation *s*- (slow) and *f*- (fast) waves have been used, whereas in the case $\theta = 0$ *r*- (right handed) and *l*- (left handed) waves are defined.

Examples for the dispersion relation according to equation (3.3.1.-4) are demonstrated in fig. 12. A transition between both modes in the regions at $X = 1$ marked by circles are only possible, if collisions are noticeably present (cf. (3.3.2.-16)).

If electron-ion collisions are taken into account, e.g. by a velocity-proportional term in the equation of motion of a single electron, the complete APPLETON-

HARTREE formula

$$n_j^2 = 1 - \frac{2\,X\,(1 - X - i\,Z)}{2\,(1 - i\,Z)\,(1 - X - i\,Z) - Y_T^2 \mp [Y_T^4 + 4\,Y_L^2\,(1 - X - i\,Z^2)]^{1/2}} \tag{3.3.1.-5}$$

is obtained, where

$$Z = \frac{\nu_{\mathrm{coll}}}{2\,\pi\,\nu}\,. \tag{3.3.1.-6}$$

ν_{coll} denotes the gas kinetic collision frequency which is used as a measure for the probability of collisions between electrons and neutral particles or ions. But

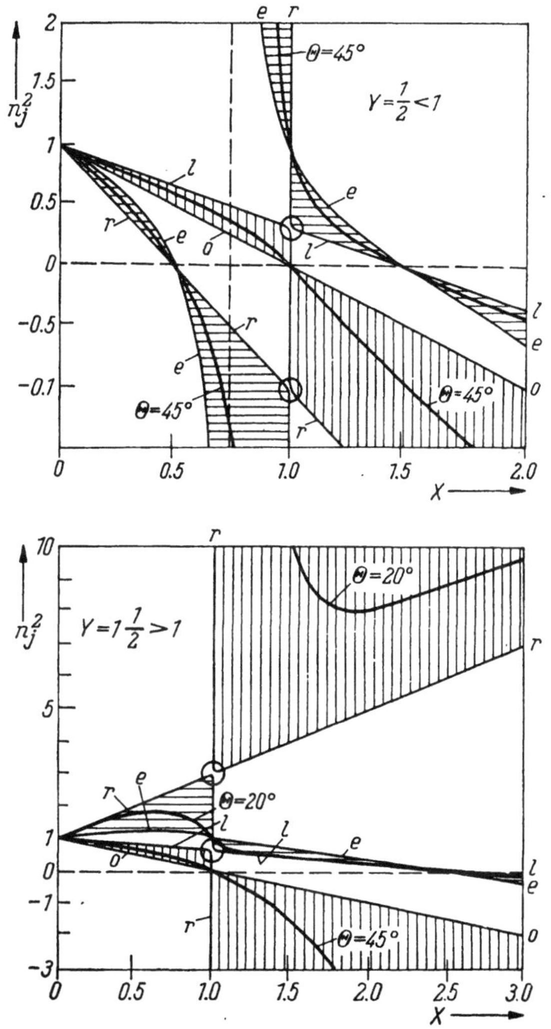

Fig. 12. Dispersion curves $n_j^2 = f(X)$ in a cold and collisionless plasma

the introduction of such a collision frequency is not free of problems (RAWER and SUCHY, 1967). In the simplest case the collision frequency is obtained as quotient of the mean (often thermal!) electron velocity and the mean free path length for collisions with a given sort of particles. Strictly speaking, the application of the collision frequency in the way used in the magneto-ionic theory is only correct, if ν_{coll} is independent of the electron velocity (cf. HEALD and WHARTON, 1965).

The following expression for ν_{coll} was given by COWLING (1945)

$$\nu_{coll} = 1.8 \cdot 10^{-8} \left(\frac{T}{300}\right)^{1/2} N_a + 6.1 \cdot 10^{-3} \left(\frac{T}{300}\right)^{-3/2} N_i \; [s^{-1}] \,. \quad (3.3.1.\text{-}7)$$

For radio astronomy, however, in most cases the first term of (3.3.1.-7) which describes collisions with the neutral gas particles can be neglected. Collisions between electrons are not contained in (3.3.1.-7).

The derivations of similar expressions for the collision frequency by other authors differ slightly from each other with respect to the definitions, prepositions, and approximations used. For instance, GINSBURG (1960) used equation (3.3.1.-5) as definition of an effective collision frequency ν_{eff}, which essentially corresponds to the form of equation (3.3.1.-7).

It must be further noted, that strictly the concept of the gas kinetic definition of the collision frequency only is applicable, if the interaction potential is no wide ranging function in space, i.e. that collision cross sections can be easily defined. Thus difficulties arise concerning the above given probability definition of the collision frequency with respect to the COULOMB interaction. In order to overcome this difficulty often a shealding radius or a shealding function is used. Other possibilities of a discussion of collision processes are given by the transport theory (RAWER and SUCHY, 1967) and by the kinetic theory (TIDMAN et al., 1966; DUPRES, 1964).

Equation (3.3.1.-5) permits to distinguish a number of special cases which are compiled in table 6.

The applicability of the quasi-longitudinal and the quasi-transversal approximations, respectively, for $X = 1$ are given by

$$\left.\begin{aligned} QL&: \frac{1}{2} Y \left|\frac{\sin 2\theta}{\cos\theta}\right| < \frac{1}{3} |(1 - X)| \\ QT&: \frac{1}{2} Y \left|\frac{\sin 2\theta}{\cos\theta}\right| > 3\, |(1 - X)| \end{aligned}\right\} \quad (3.3.1.\text{-}8)$$

if the inequality in (3.3.1.-8) does not exceed the ratio 1:9 (RATCLIFFE, 1959).

The singularities and zeros of the refractive index describing the resonance and cut-off characteristics are of special interest. This corresponds to the existence of certain "critical" frequencies indicating the regions where in the approximation of the magneto-ionic theory a propagation of waves is not possible. For an investigation of the propagation characteristics at these critical points themselves, however, the theory of the cold plasma as well as geometrical optics are hardly to be applied so that other methods of wave optics must be used.

Tab. 6. Special cases of the dispersion relation for electromagnetic wave modes

	$Y = 0$	$Y_T = 0\ (\omega_H \sin\theta \ll \omega)$	$Y_L = 0\ (\omega_H \cos\theta \ll \omega)$
$Z = 0$ $\left(\nu_{coll} \ll \frac{\omega}{2\pi}\right)$	$n^2 = 1 - X$	$n_j^2 = 1 - \frac{X}{1 \mp Y_L}$	$n_1^2 = 1 - \frac{X(1-X)}{1 - X - Y_T^2}$ $n_2^2 = 1 - X$
$Z \neq 0$	$n^2 = 1 - \frac{X}{1 - iZ}$	$n_j^2 = 1 - \frac{X}{1 \mp Y_L - iZ}$	$n_1^2 = 1 - \frac{X(1 - X - iZ)}{(1 - iZ)(1 - X - iZ) - Y_T^2}$ $n_2^2 = 1 - \frac{X}{1 - iZ}$
	$X = 0$ $(\omega_p \ll \omega)$	$QL: \frac{Y_t^4}{4 Y_T^2} \ll \lvert(1 - X - iZ)^2\rvert$	$QT: \lvert(1 - X - iZ)^2\rvert \ll \frac{Y_T^4}{4 Y_L^2}$
	$n = 1$	$n_j^2 = 1 - \frac{X}{1 \mp Y_L - iZ}$	$n_1^2 = 1 - \frac{X(1 - X - iZ)}{(1 - iZ)(1 - X - iZ) - Y_T^2}$ $n_2^2 = 1 - \frac{X}{1 - iZ + (1 - X - iZ)\, Y_L^2/Y_T^2}$

According to (3.3.1.-5) for arbitrary angles θ in general three zeros of the refractive index exist which are given in table 7. If $\theta = 0$, the number of zeros is reduced to two, and for $Y = 0$ only one zero is present.

If $Z = 0$, X becomes complex satisfying the condition $n^2 = 0$.

Similarily, the poles of the refractive are obtained from (3.3.1.-5) as follows

$$Y = \left(\frac{1 - X}{1 - X \cos^2 \theta}\right)^{1/2}, \quad X = \frac{1 - Y^2}{1 - Y^2 \cos^2 \theta} \quad (Z = 0) \tag{3.3.1.-9}$$

or

$$Y = \left(\frac{(1 - i\,Z)\,(1 - X - i\,Z)}{1 - i\,Z - X \cos^2 \theta}\right)^{1/2}, \quad X = (1 - i\,Z)\frac{(1 - i\,Z)^2 - Y^2}{(1 - i\,Z)^2 - Y^2 \cos^2 \theta} \quad (Z \neq 0)\,. \tag{3.3.1.-10}$$

From these equations the well known singularities e.g. at the gyrofrequency for $Z = 0$ and propagation in direction of the magnetic field, and at the hybrid frequency ("magnetic plasma frequency") follow (cf. table 7). Resonances at harmonics of the gyrofrequency are not contained in the approximation of the cold plasma. These effects are obtained by calculation basing of the kinetic theory (cf. section 3.4).

Besides fig. 12 different kinds of representations of dispersion relations are possible. Especially contour maps of n_j and polar plots of $\frac{1}{n_j} = \frac{v_{k_j}}{c}$ (phase velocity or wave normal surfaces) are instructive. A detailed treatment on this subject is contained in HEALD and WHARTON (1965). A concise representation of the topology of the wave normal surfaces is further given in a schematized form by the CMA-diagrams named after CLEMMOW, MULLALY, and ALLIS (cf. e.g. STIX, 1962). Such diagrams provide a synoptic picture of the shapes of the wave normal surfaces, the resonances, and cutoffs in different parameter regions not only in the cold plasma but especially also for the wave modes of the warm plasma (fig. 13).

3.3.2. Polarization characteristics of electromagnetic waves

In the foregoing sections only the intensity or flux density of the radiation was considered without regard to the polarization characteristics of radio waves. In general, due to the influence of external magnetic fields, however, the solar radio emission is composed of a polarized and a non-polarized (i.e. randomly polarized) component:

$$I_\nu = I_{\nu,\mathrm{pol}} + I_{\nu,\mathrm{ran}}\,. \tag{3.3.2.-1}$$

The degree of polarization is defined as the ratio between the polarized part and the total radiation as follows

$$\varrho = \frac{I_{\nu,\mathrm{pol}}}{I_\nu}\,. \tag{3.3.2.-2}$$

Tab. 7. Zeros and poles of the refractive index

	critical frequencies $n_j = 0$	resonances $(n_j = \infty)$ $\theta = 0$	$\theta = \frac{\pi}{2}$
$Z = 0$	$X = 1; \quad \omega = \omega_p; \quad \omega_H = 0$ $X = 1 \mp Y$ $\omega_j = \sqrt{\omega_p^2 + \frac{\omega_H^2}{4}} \pm \frac{\omega_H}{2}$	$[X = 1; \omega = \omega_p]$ $Y_L = 1; \omega = \omega_H$	$X = 1$ $X = 1 - Y_T^2$ $\omega_M = \sqrt{\omega_p^2 + \omega_H^2}$
$Z \neq 0$	$X = 1 - i Z$ $\omega_j = \frac{i \nu_{coll}}{2} \pm \sqrt{\omega_p^2 - \frac{\nu_{coll}^2}{4}}$ $X = 1 - i Z \pm Y$ $\omega_j = \frac{i \nu_{coll} \mp \omega_H}{2} \pm$ $\pm \sqrt{\omega_p^2 + \frac{i \nu_{coll} \mp \omega_H^2}{4}}$	$X = 1 - i Z$ $\omega_j = \frac{i \nu_{coll}}{2} \pm \sqrt{\omega_p^2 - \frac{\nu_{coll}^2}{4}}$	$X = \frac{(1 - i Z)^2 - Y_T^2}{1 - i Z}$ $\omega_j = i \nu_{coll} \pm \sqrt{\omega_p^2 + \omega_H^2 - \nu_{coll}^2}$ $(Z \ll 1)$

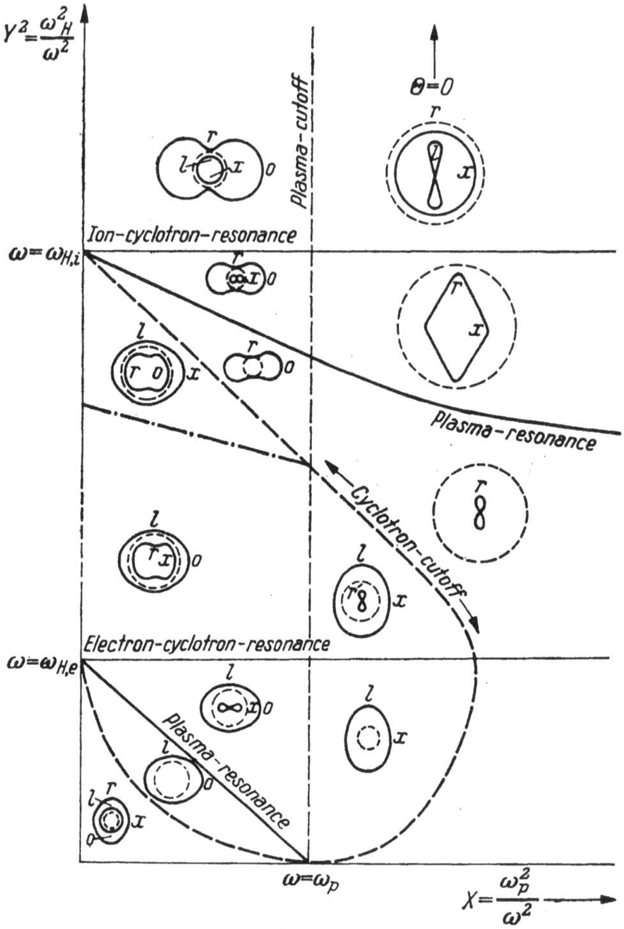

Fig. 13. CMA diagram for a two component plasma

The vector of the electrical field of a polarized electromagnetic wave perpendicular to the direction of propagation consists of following components

$$\left.\begin{aligned} E_{x,\mathrm{pol}}(t) &= E_{0x} \sin(\omega_0 t - \varphi_x) \\ E_{y,\mathrm{pol}}(t) &= E_{0y} \sin(\omega_0 t - \varphi_y) \,. \end{aligned}\right\} \qquad (3.3.2.\text{-}3)$$

These components define an ellipse, for which the amplitudes E_{0x}, E_{0y} determine the axial ratio p'. The phase difference $(\varphi_x - \varphi_y) = \varphi_{xy}$ is given by the orientation of the polarization ellipse expressed by the angle χ between the main axis of the ellipse and the x-axis of a Cartesian coordinate system. Altogether there are four quantities I_v, ϱ, p, and χ which describe the polarization characteristics of an electromagnetic wave. These quantities can be easily transformed into the widely used STOKES parameters I, Q, V, and U

$$\left.\begin{aligned} I_\nu &= I \\ \varrho &= \frac{(Q^2 + U^2 + V^2)^{1/2}}{I} \\ \sin(2\sigma) &= \frac{V}{I_{\nu,\mathrm{pol}}} = \frac{V}{(Q^2 + U^2 + V^2)^{1/2}} \\ \operatorname{tg} 2\chi &= \frac{U}{Q} \qquad \text{where} \qquad p' = \operatorname{tg}\sigma\,. \end{aligned}\right\} \tag{3.3.2.-4}$$

For the STOKES parameters the following relations are valid

$$\begin{aligned} I &= I_\nu = I_{\nu,x} + I_{\nu,y} \\ Q &= \varrho I_\nu \cos(2\sigma)\cos(2\chi) = I_{\nu,x} - I_{\nu,y} \\ U &= \varrho I_\nu \cos(2\sigma)\sin(2\chi) = \overline{2 E_{0x} E_{0y} \cos\varphi_{xy}} \\ V &= \varrho I_\nu \sin(2\sigma) = \overline{2 E_{0x} E_{0y}} \sin\varphi_{xy}\,, \end{aligned} \tag{3.3.2.-5}$$

whereas

$$\begin{aligned} I_{\nu,x} &= I_{\nu,x,\mathrm{pol}} + \frac{I_{\nu,\mathrm{ran}}}{2} \\ I_{\nu,y} &= I_{\nu,y,\mathrm{pol}} + \frac{I_{\nu,\mathrm{ran}}}{2} \end{aligned} \tag{3.3.2.-6}$$

are the linear components of the radiation in x- and y-direction, respectively.

Furthermore we have

$$I_{\nu,\mathrm{pol}} = \overline{E_{0x}^2} + \overline{E_{0y}^2} = I_{\nu,x,\mathrm{pol}} + I_{\nu,y,\mathrm{pol}} \tag{3.3.2.-7}$$

when the averaging over the time is carried out.

If pure linear polarized radiation is considered, the degree of polarization is

$$\varrho_l = \frac{I_{\nu,x} - I_{\nu,y}}{(I_{\nu,x} + I_{\nu,y})\cos(2\chi)} = \frac{Q}{I\cos(2\chi)} = \varrho\cos(2\sigma)\,. \tag{3.3.2.-8}$$

Analogeously for circular polarization

$$\varrho_c = \frac{V}{I} = \varrho\sin(2\sigma) \tag{3.3.2.-9}$$

is obtained.

In general it is valid

$$\varrho = (\varrho_l^2 + \varrho_z^2)^{1/2} \tag{3.3.2.-10}$$

If an elliptically polarized wave is propagating in a magnetoactive plasma, due to the effect of anisotropy all components of the field vectors (E_x, E_y, E_z, H_x, H_y, H_z) [1]) are existing.

[1]) The z-axis is in direction of the wave propagation, the x-axis denotes the direction of the magnetic field component perpendicular to the direction of the propagation.

According to RATCLIFFE (1959) the polarization is described by the complex ratios

$$\frac{E_y}{E_x}, \quad \frac{E_z}{E_x}, \quad \frac{H_y}{H_x}, \quad \frac{H_z}{H_x}.$$

In direction of wave propagation these ratios are only constant for two certain characteristic waves which satisfy the condition

$$\frac{P_x}{E_x} = \frac{P_y}{E_y} \tag{3.3.2.-11}$$

(P_x, P_y are the components of the dielectric polarization vector). For both characteristic waves the axis ratio of the polarization ellipses is determined by the expression

$$R = \frac{i}{\operatorname{tg} \sigma} = \frac{E_x}{E_y} = -\frac{H_y}{H_x} = \frac{P_x}{P_y} \tag{3.3.2.-12}$$

or, explicitely written

$$\begin{aligned} R_j &= -i\left\{\frac{Y_T^2}{2\,Y_L(1 - X - i\,Z)} \mp \left[\frac{Y_T^4}{4\,Y_L^2(1 - X - i\,Z)^2} + 1\right]^{1/2}\right\} \\ &= -\frac{i}{Y\cos\theta}\left\{\frac{Y^2\sin^2\theta}{2\,(1 - x - i\,Z)} \mp \left[\frac{Y^4\sin^4\theta}{4\,(1 - X - i\,Z)} + Y^2\cos^2\theta\right]^{1/2}\right\}. \end{aligned} \tag{3.3.2.-13}$$

Sometimes, instead of R_j also the quantity K_j is used:

$$K_j = \frac{i}{R_j} = \frac{2\,Y_L(1 - X - i\,Z)}{Y_T^2 \mp [Y_T^4 + 4\,Y_L^4\,(1 - X - i\,Z)]^{1/2}} = \eta \mp (\eta^2 + 1)^{1/2}, \tag{3.3.2.-14}$$

whereas

$$\eta = \frac{Y}{(1 - X - i\,Z)}\,\frac{\sin^2\theta}{2\cos\theta}. \tag{3.3.2.-15}$$

Because $R_1R_2 = 1$, both polarization ellipses exhibit the same form but having inverse orientations. For $Z = 0$, R_j becomes pure imaginary. Then the main axes of the ellipses are parallel to the coordinate axis perpendicular to the direction of wave propagation. In the general case $Z \neq 0$ both polarization ellipses are placed symmetrically to a straight line of 45° between the x- and y-axis.

At a "critical collision frequency" Z_0

$$Z_0 = \left|\frac{Y_T^2}{2\,Y_L}\right| \quad \text{with} \quad X = 1 \tag{3.3.2.-16}$$

both ellipses degenerate into a straight line. In (3.3.2.-16) the angle θ is contained. Therefore the existence of a critical collision frequency corresponds to the occurrence of a critical angle θ_0, which is given by

$$\frac{\sin^2\theta}{\cos\theta} = \frac{2\,\nu_{\text{coll}}}{\omega_H} \quad (X = 1). \tag{3.3.2.-17}$$

For

$\theta > \theta_0$ there is at $X = 1$ a continuous transition of changing the sense of polarization within one and the same mode,

$\theta = \theta_0$ there takes place a coupling of the characteristic waves,

$\theta < \theta_0$ there is a change of modes, but no change of the sense of polarization when passing the point $X = 1$.

If the propagation is perpendicular to the direction of the magnetic field, linear polarization is obtained ($R_j = 0, \infty$); the vector of the E-field of the ordinary wave component is parallel to the external magnetic field and perpendicular to the E-vector of the extraordinary component.

For propagation parallel to the magnetic field direction ($R_j = \pm i$) we have a circular polarization of the characteristic waves. The sense of orientation of the extraordinary component corresponds to that of the gyration of the electrons in a magnetic field. These properties of R_j in dependence of the angle θ and the notation of the wave modes according to Stix (1962) are compiled in table 8.

The geometry of left and right handed polarized waves with respect to their definitions according to the direction of the wave propagation or magnetic field is shown in fig. 14.

For propagation in an inhomogeneous medium with varying parameters X, Y, Z in general the polarization of the characteristic waves is changed. But according to (3.3.2.-13) the polarization remains constant, if $Y = 0$. This case is important for the estimation of the limiting polarization of the radiation coming from the Sun, which for $\theta \neq \frac{\pi}{2}$ should be circular. The orientation should depend on the angle θ (acute or obtuse) when the radiation leaves the magnetic field. But it must be mentioned, that for the transition $Y \to 0$ (analogously to $X \to 0$) the simple calculation of geometric optics is not longer justified. Instead of this a coupling between the ordinary and extraordinary waves can occur influencing the state of polarization. Therefore quantitative conclusions are only possible in connection with concrete models.

For $X \to 0$ one obtains from (3.3.2.-13)

$$R_j = -i\left[\frac{Y_T^2}{2\,Y_L} \mp \left(\frac{Y_T^4}{4\,Y_L^2} + 1\right)^{1/2}\right]. \tag{3.3.2.-18}$$

Tab. 8. Polarization and wave modes in a cold plasma

θ	Modes	R_j	Polarization
0	r, l	$\pm i$	circular
$0 \cdots \frac{\pi}{2}$	s, f	complex	elliptical
$\frac{\pi}{2}$	$e(x), 0$	real	linear

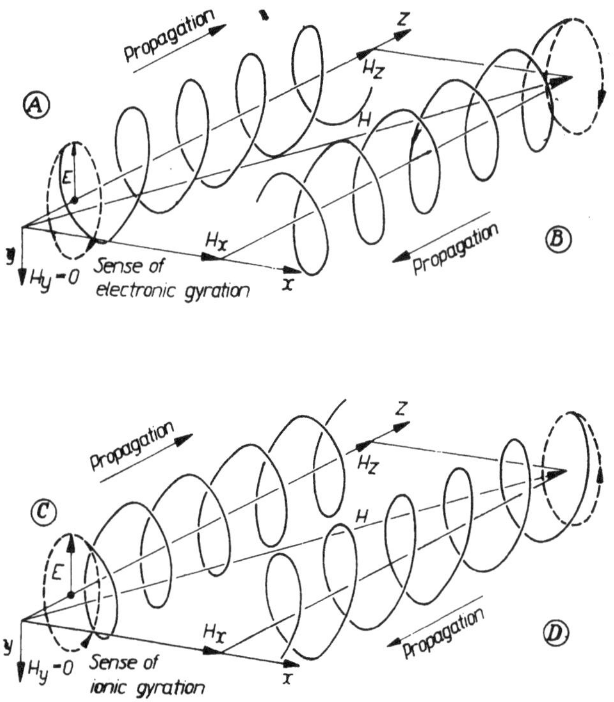

Case	Propagation	Gyration	Space dependence	Polarization Def. a	(b)	c
A	$\parallel H_z$	electronic	left	right	(left)	right
B	$\uparrow\downarrow H_z$	electronic	right	left	(right)	right
C	$\parallel H_z$	ionic	right	left	(right)	left
D	$\uparrow\downarrow H_z$	ionic	left	right	(left)	left

Fig. 14. Schematic representation of the geometry of right and left handed polarized waves with respect to different definitions:

a – Definition commonly used in radio astronomy (observer looks in direction of the wave propagation)

b – Definition commonly used in crystal optics (observer looks at the source)

c – Definition used in laboratory plasma physics regarding the sense of gyration looking in direction of the magnetic field H_z

From this, even in inhomogeneous magnetic fields, one could expect no changes of polarization. But really this case can be considered as limiting case $\nu_H \gg \nu_p$ only. If the particle density is exactly vanishing, the derivation of (3.3.2.-13) is no longer valid which was based on the fundamental equations of the kinetic theory or the equation of motion of electrons. In fact, in the vacuum a change of the state of polarization of course is not possible as it directly follows from the MAXWELL equations.

A special case is obtained, when $X = 1$. Here also the use of (3.3.2.-13) is possible with caution only, which for arbitrary angles θ yields linear polarization.

The properties of R_j can be easily derived from a diagram proposed by MOLLWO (1969). For this purpose equation (3.3.2.-13) is written in the form

$$(\text{sign}\cos\theta)\,R_j = \frac{-iZ}{1-X-iZ}\left[\frac{Y\sin^2\theta}{2Z|\cos\theta|} \mp (\text{sign}\cos\theta)\left\{\left(\frac{Y\sin^2\theta}{2Z|\cos\theta|}\right)^2 + \right.\right.$$
$$\left.\left. + \left(\frac{1-X-iZ}{Z}\right)\right\}^2\right]. \qquad (3.3.2.\text{-}13\text{a})$$

Then it follows from the dispersion relation (3.3.1.-4), that the real part of $(\text{sign}\cos\theta)\,R_j$ must be always positive and a conform mapping into the complex half plane becomes possible. This mapping of the positive half plane of a rectangular coordinate system into the unit circle by the function

$$w = u + i\,v = \frac{(\text{sign}\cos\theta)\,R_j - 1}{(\text{sign}\cos\theta)\,R_j + 1}$$

is shown in fig. 15. There are drawn curves of the parameters $\frac{Y}{Z}\frac{\sin^2\theta}{2\cos\theta} =$ $=$ const and $\frac{1-X}{Z} =$ const. The critical point given by (3.3.2.-16) is represented by the centre of the diagram of fig. 15. For all other combinations of the plasma parameters X, Y, Z two points exist which are located symmetrical to the centre of the diagram corresponding to the polarizations of the two characteristic waves ($R_1 R_2 = 1$). The case $Z = 0$ is represented by the periphery of the unit circle and only one parameter $\frac{Y}{1-X}\frac{\sin^2\theta}{2\cos\theta}$ is needed for the description of the state of polarization. It should be mentioned, that for $Z \neq 0$ this parameter and the quantity $\frac{Y}{Z}\frac{\sin^2\theta}{\cos\theta}$ define a set of orthogonal curves in the complex R-plane (SNYDER and HELLIWELL, 1952), which can be used also for a representation of the wave polarization similar to fig. 15. The special form of fig. 15 was choosen for practical reasons.

For greater anisotropies farther the ratio

$$\frac{E_z}{E_x} = G_j = i\,\Gamma_j\,, \qquad (3.3.2.\text{-}19)$$

is defined whereas

$$G_j = \frac{i\,X\,Y_T}{(1-X-iZ)\left\{\left[1-iZ-\frac{Y_T^2}{2(1-X-iZ)}\right] \pm \left[\frac{Y_T^4}{4(1-X-iZ)^2} + Y_L^2\right]^{1/2}\right\}}$$
$$= \frac{i\,Y_T\,(n_j^2-1)}{1-X-iZ}\,. \qquad (3.3.2.\text{-}20)$$

Concerning the degree of polarization the following remarks can be given. In general a remarkable amount of polarized radiation results, if the two charac-

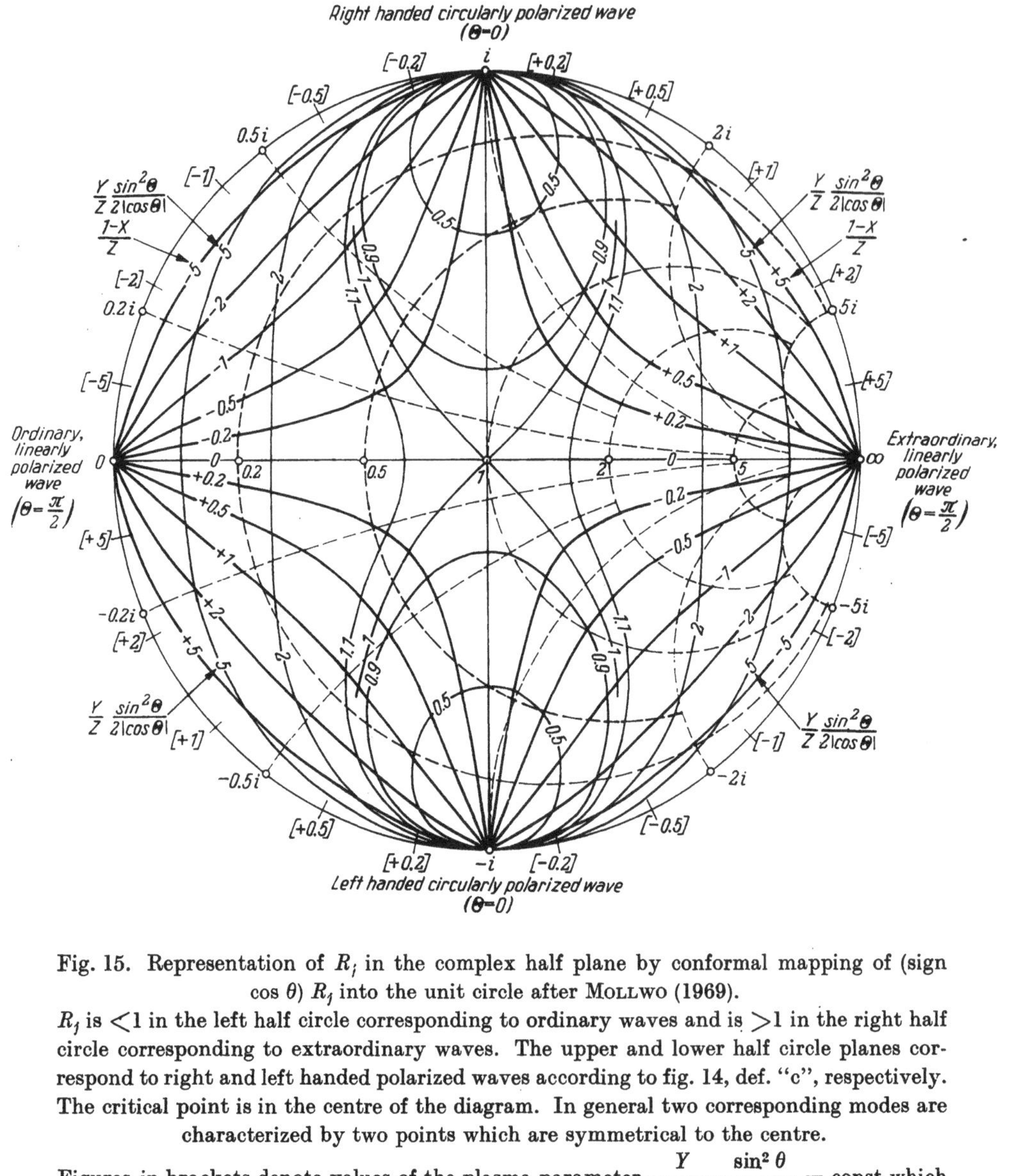

Fig. 15. Representation of R_j in the complex half plane by conformal mapping of (sign $\cos\theta$) R_j into the unit circle after MOLLWO (1969).

R_j is <1 in the left half circle corresponding to ordinary waves and is >1 in the right half circle corresponding to extraordinary waves. The upper and lower half circle planes correspond to right and left handed polarized waves according to fig. 14, def. "c", respectively. The critical point is in the centre of the diagram. In general two corresponding modes are characterized by two points which are symmetrical to the centre.

Figures in brackets denote values of the plasma parameter $\frac{Y}{1-X}\,\frac{\sin^2\theta}{2\,|\cos\theta|} = \text{const}$ which define orthogonal trajectories to the isolines of the parameter $\frac{Y}{Z}\,\frac{\sin^2\theta}{2\,|\cos\theta|}$ (cf. SNYDER and HELLIWELL, 1952)

teristic waves are not of the same intensity. Hence, for an estimation of the degree of polarization different conditions of the generation of the ordinary and extraordinary wave modes, especially in the neighbourhood of critical frequencies are important. For instance, the emission of the extraordinary component near the level $Y = 1 - X$ causes a predominance of the polarization of this component because of the shielding of the ordinary component in that region. Calculations of special models are needed in order to get more detailed conclusions. In regions far from critical frequencies, however, the propagation of the radiation of the ordinary and extraordinary components for both modes is approximately the same and does not influence the polarization by selective effects.

At last the position of the polarization ellipses must be taken into consideration. The polarized radiation can be considered as consisting of two coherent polarized components corresponding to the magneto-ionic wave modes, which may be written as

$$\left.\begin{aligned} E_{x,1} &= A \sin(\omega t)\,, & E_{y,1} &= A\, p' \cos(\omega t) \\ E_{x,2} &= B p' \sin(\omega t - \varphi)\,, & E_{y,2} &= B \cos(\omega t - \varphi)\,. \end{aligned}\right\} \quad (3.3.2.\text{-}21)$$

For the character of the polarization the phase angle φ and the axis ratio p' are fundamental quantities.

The position of the main axes of the resulting polarization ellipse is obtained from the following relation

$$\operatorname{tg} 2\chi = \frac{-\dfrac{1+p'^2}{p'}\, 2 \sin\varphi}{\left(\dfrac{1}{p'} - p'\right)\left(\dfrac{A}{B} - \dfrac{B}{A}\right) + 4\cos\varphi} \quad (3.3.2.\text{-}22)$$

(cf. ZHELEZNYAKOV, 1964).

Passing an anisotropic medium, the phase angle is changed by

$$\Delta\varphi = \frac{\omega}{c} \int_0^s (n_2 - n_1)\, ds \quad (3.3.2.\text{-}23)$$

(FARADAY effect).

In the absence of collisions for the case of the QL-approximation it is obtained

$$\Delta\chi = \frac{\Delta\varphi}{2} \approx \frac{1}{\omega^2 c} \int_0^s \frac{\omega_p^2\, \omega_H \cos\theta}{n_1 + n_2}\, ds\,. \quad (3.3.2.\text{-}24)$$

Hence a large influence of the FARADAY rotation on the propagation of radio waves in the solar atmosphere follows. Assuming, e.g., $N_e = 10^{10} \cdots 10^8\ \mathrm{cm}^{-3}$, $H \approx 20$ gauss, and $s = 10^{10}$ cm, $\Delta\chi$ is roughly estimated to

$$\Delta\chi \approx (10^{26} \cdots 20^{24})\, \nu^{-2}\,. \quad (3.3.2.\text{-}25)$$

If finite ray path differences $\Delta s$are considered, from (3.3.2.-24) a local $\delta\chi$ is obtained which is of the order of

$$\delta\chi \approx 4.7 \cdot 10^4 \nu^{-2} N_e H \cos\theta \Delta s . \quad (3.3.2.\text{-}26)$$

In the case

$$\delta\chi > 1 \quad (3.3.2.\text{-}27)$$

inside the emission regions a large scattering of the polarization ellipses can be expected. This leads to a "depolarization effect" of the linear polarized component, i.e. to a conversion of elliptical into more and more circular polarization. Then the condition (3.3.2.-27) together with (3.3.2.-26) determine an upper limit of the characteristical dimension of the emission regions, if elliptical polarization is observed. This limit for the cm-region is at best of the order of 10 km, which for bursts in most cases should be exceeded. Of course, the above given estimations are very roughly and made only for the case of the QL-approximation under the assumption that both, the ordinary and the extraordinary waves are emitted and propagated. Occasionally observed components of partly linear polarized burst radiations lead to the supposition that the above made prepositions need not necessarily be fulfilled in all cases of burst events.

An instrumental effect of a similar "depolarization" of linear polarization is due to the finite bandwidth of the receivers (FARADAY dispersion). Its influence can be estimated e.g. again for the QL-case under the simplifying assumption $n_1 \approx 1$, which according to (3.3.2.-24) leads to

$$\delta\chi \approx \frac{2}{\nu_r} \Delta\chi_{\nu=\nu_r} \Delta\nu_r \quad (3.3.2.\text{-}28)$$

(ν_r — observing frequency, $\Delta\nu_r$ — bandwidth of the receiver).

3.3.3. *Coupling of electromagnetic wave modes in a cold plasma*

The statements of the previous sections refer to local homogeniety of the medium in which the wave propagation takes place. If stronger inhomogenieties are present, the theory of coupling of the wave modes must be taken into account. Here especially the coupling between the ordinary and the extraordinary waves within the framework of the magneto-ionic theory excluding the point $X = 1$ will be considered. Other possible coupling processes e.g. due to effects in a warm plasma or non-linear wave interactions will be touched lateron.

The first attempts of a theory of mode coupling were made with respect to applications to the physics of the ionosphere (see e.g. RATCLIFFE, 1959; BUDDEN, 1961). For applications in solar physics, however, some generalizations, in particular the extension to variable magnetic fields, are necessary.

Following to COHEN (1960) at first we consider a medium which is stratified in plane layers perpendicular to the direction of wave propagation. Starting from the coupled wave equations

$$\left.\begin{aligned} \Pi_1'' + n_1^2 \Pi_1 &= -\Psi_1 \Psi_2 \Pi_1 + \Psi_2' \Pi_2 + 2\Psi_2 \Pi_2' \\ \Pi_2'' + n_2^2 \Pi_2 &= -\Psi_2 \Psi_1 \Pi_2 + \Psi_1' \Pi_1 + 2\Psi_1 \Pi_1' \end{aligned}\right\} \quad (3.3.3.\text{-}1)$$

with

$$E_{y,1} = \Pi_1 \exp\left(-\int_0^s \frac{R_1}{R_1 - R_2} ds\right); \quad E_{y,2} = \Pi_2 \exp\left(-\int_0^s \frac{R_2}{R_2 - R_1} ds\right) \tag{3.3.3.-2}$$

the coupling parameters are written as

$$\left.\begin{aligned} \Psi_1 &= \frac{R_1'}{R_1 - R_2} \exp\left(\int_0^s \frac{R_2' + R_1'}{R_2 - R_1} ds\right) \\ \Psi_2 &= \frac{R_2'}{R_2 - R_1} \exp\left(\int_0^s \frac{R_1' + R_2'}{R_1 - R_2} ds\right) \end{aligned}\right\} \tag{3.3.3.-3}$$

(′ denote derivations with respect to the ray path s).

In this representation the theory of mode coupling originally developed in connection with ionospheric problems (FÖRSTERLING, 1942; RYDBECK, 1944; BUDDEN, 1961) is contained as special case if $Z = 0$ and only one coupling parameter $|\Psi| = |\Psi_1| = |\Psi_2|$ is present.

The system of differential equations (3.3.3.-1) can be solved approximately by using the WKB-method. With the statement

$$\Pi_j = U_j + \exp\left\{-\frac{i}{2}\int_0^s (n_1 + n_2)\, ds\right\} \tag{3.3.3.-4}$$

and under the assumption $X, Y, Z \ll 1$ the following simplified system of the coupled differential equations is obtained

$$\left.\begin{aligned} \frac{dU_1}{ds} &= \Psi_2 U_2 + \frac{i}{2}(n_2 - n_1) U_1 \\ \frac{dU_2}{ds} &= \Psi_1 U_1 + \frac{i}{2}(n_1 - n_2) U_2 . \end{aligned}\right\} \tag{3.3.3.-5}$$

From this it is seen, that the degree of coupling is described by the ratios

$$\frac{\left|\frac{i}{2}(n_1 - n_2)\right|}{|\Psi_1|}, \quad \frac{\left|\frac{i}{2}(n_2 - n_1)\right|}{|\Psi_2|}.$$

Therefore it appears convenient to define the expression

$$Q' = \left(\frac{|\Psi_1 \Psi_2|}{\frac{1}{4}(n_2 - n_1)^2}\right)^{1/2} \tag{3.3.3.-6}$$

as coupling ratio.

Now the limiting case of weak mode coupling may be considered

$$\left|\frac{i}{2}(n_2 - n_1)\right| \gg |\Psi_j|, \qquad Q' \ll 1. \tag{3.3.3.-7}$$

In this case the first term of the right side in both equations of (3.3.3.-5) can be neglected. Then the solutions are

$$U_j = \exp\left\{\pm \frac{i}{2} \int_0^s (n_2 - n_1)\, ds\right\} \tag{3.3.3.-8}$$

and the wave functions (3.3.3.-4) are written in the form

$$\Pi_j = \exp\left\{- i\, k_j \int_0^s n_j\, ds\right\} \quad \text{where } s = k_j\, z\,. \tag{3.3.3.-9}$$

The resulting wave propagates as the sum of the extraordinary and the ordinary components which are independent of each other, i.e. no coupling of wave modes exist. All propagation characteristics including the state of polarization are closely related to the actual physical condition of the medium passed by the waves. This is equivalent to the case of the local homogeniety firstly described and discussed by BOOKER (1936).

The other extreme case referring to strong coupling is represented by

$$\left|\frac{i}{2}(n_2 - n_1)\right| \ll \Psi_j\,, \qquad Q' \gg 1\,. \tag{3.3.3.-10}$$

Here the solution of the system of differential equations (3.3.3.-5) is given by the linear combinations of solutions which have the following form

$$U_{11} = -\exp\left(\int_0^s \frac{R_2'}{R_1 - R_2}\, ds\right), \qquad U_{21} = \exp\left(\int_0^s \frac{R_1'}{R_2 - R_1}\, ds\right)$$

$$U_{12} = \frac{R_2}{R_2 - R_1} \exp\left(\int_0^s \frac{R_1'}{R_1 - R_2}\, ds\right), \qquad U_{22} = \frac{R_1''}{R_1 - R_2}\left(\exp \int_0^2 \frac{R_2'}{R_2 - R_1}\, ds\right). \tag{3.3.3.-11}$$

Because the solutions (3.3.3.-11) contain only one $\bar{n} = \frac{1}{2}(n_1 + n_2)$, in this case both modes are coupled together and propagating with equal propagation characteristics, which can be concluded also directly from (3.3.3.-9). Therefore, in the case of strong coupling, such effects like FARADAY rotation or differential absorption of single wave modes are not possible. It should be noted, that under the condition X, Y, $Z \ll 1$, the propagation takes place without reference to the outer magnetic field, because $\bar{n}$ is identical with the refractive index for the case without external magnetic fields.

Now the case of arbitrary Q' is to be considered. Then, again after COHEN (1960), under the above made assumptions for the case of QL-approximation it follows

$$Q'_{\mathrm{QL}} = \frac{C}{\nu^2\, X\, Y \cos\theta}\,. \tag{3.3.3.-12}$$

The case $Q' = 1$ is to be regarded as the transition between the cases of weak and strong coupling. Thus, with a given geometry of the medium (which is contained in the term C of (3.3.3.-12)) and given X, Y, and θ from $Q' = 1$ a special transitional frequency ν_t can be estimated, so that for $\nu \gg \nu_t$ there is a strong coupling, and for $\nu \ll \nu_t$ there is a weak coupling of the wave modes, respectively.

Cohen (1960) has estimated the magnitudes of ν_t for several typical cases of different plasma parameters which are shown in a slightly completed form in table 9. From this it follows, that in the QL-approximation in general the transitional frequencies are far above the region of radio frequencies. Thus, in all QL-regions only a weak coupling of the magneto-ionic wave modes may be expected.

Tab. 9. Values of the transitional frequency ν_t in different media

	QL	QT
Interstellar space	10^5 GHz	50 kHz
Solar Corona ($r = 2\,R_\odot$)	$5 \cdot 10^4$ GHz	$5 \cdot 10^2$ MHz
Interplanetary medium	$5 \cdot 10^3$ GHz	0.1 MHz
Ionosphere (F_2-layer)	10^3 GHz	50 MHz

Other conditions are found for the QT-approximation. Here it is obtained

$$Q'_{\mathrm{QT}} \sim \frac{\nu^4}{N_e \cdot h' \cdot H^3}\,. \tag{3.3.3.-13}$$

(h'—scale of the magnetic field). Hence it follows

$$\nu^4_{t,\mathrm{QT}} = 10^{17}\, N_{e\,[\mathrm{cm}^{-3}]} \cdot h'_{[\mathrm{cm}]} \cdot H^3_{[\mathrm{gauss}]}\ [\mathrm{s}^{-4}]\,. \tag{3.3.3.-14}$$

Some values of ν_t calculated according to (3.3.3.-14) are given in table 9. It is easily seen, that for the QT-regions the transitional frequencies are much smaller than those of the QL-regions. Often they are even below the corresponding plasma level. Therefore, in such cases the influence of the magneto-ionic mode coupling is expected to be important. A schematical representation of the positions of the Ql- and QT-regions for bipolar and unipolar spot groups is proposed in fig. 16, but it should be additionally taken into mind, that according to the condition

$$\left|\left\{\frac{Y\dfrac{\sin^2\theta}{\cos\theta}}{2\,(1 - X - i\,Z)}\right\}^2\right| \begin{matrix} \gg 1 & \text{for QT approximation} \\ \ll 1 & \text{for QL approximation} \end{matrix} \tag{3.3.3.-15}$$

(cf. Heald and Wharton, 1965) the QL-regions are unlikely more extended than the regions of the validity of the QT-approximation.

The effect of the inversion of the orientation of the polarized radiation at the transitional frequency ν_t has been applied for an explanation of the often observed reversal of the sense of circular polarization in the spectrum of microwave

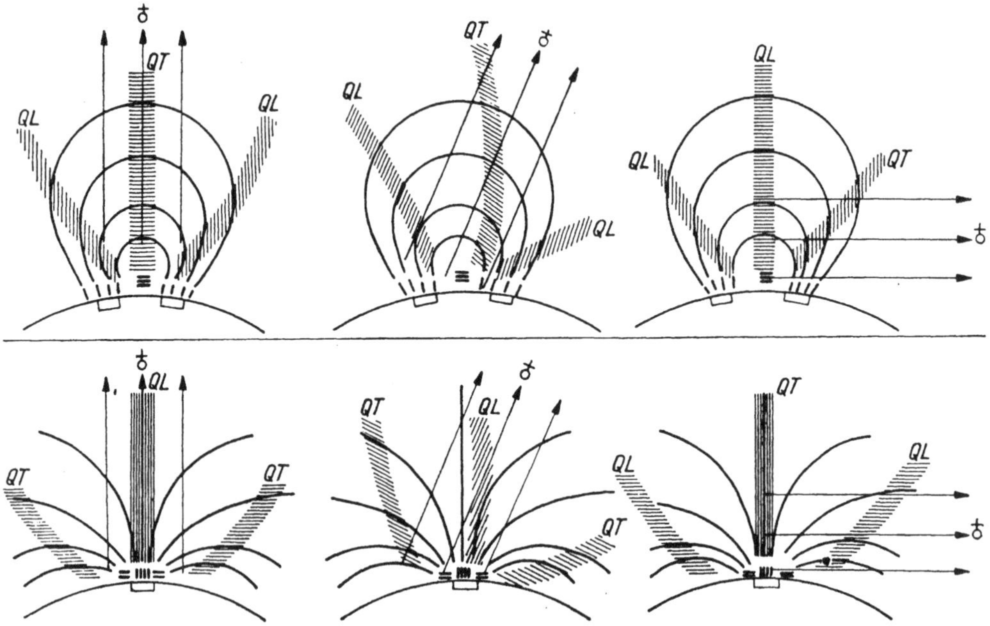

Fig. 16. Schematic representation of QL and QT regions above bipolar and unipolar sunspot regions

bursts. Ist must be assumed that the burst radiation is emitted by anyone mechanism causing a predominance of one wave mode. All frequencies $\nu > \nu_t$ exhibit a close coupling of the both wave modes and the passage of a QT-region does not influence the state of polarization. For $\nu < \nu_t$, on the other hand, an inversion of the orientation of the polarization is observed. If, due to special field configurations, several QT-regions are passed, this process can be repeated and also several inversions of the orientation of polarization would be observed.

According to (3.3.3.-14), under reasonable assumptions for N_e and h', from the position of the transitional frequency ν_t principially the strength of the magnetic field in the reversing region can be stimated. For example, if [$\nu_t = 3000$ MHz, $N_e = 10^8$ cm^{-3} and $h' = 10^{10}$ cm, a magnetic field of the order of 10 gauss would be obtained. At the transitional frequency itself the polarization should be linear. But it appears difficult to measure this effect, because due to the depolarization effects the linear polarized radiation is expected to be detectable only within small bandwidths (10$\cdots$100 kHz).

Though the above mentioned effect of magneto-ionic mode coupling leads to a plausible interpretation of the reversal of the sense of polarization of microwave bursts, the following difficulty must be noted. As a consequence of the height dependency of the transitional frequency in the solar atmosphere, the value of ν_t changes with variable direction to the observer, i.e. heliographic longitude of the burst source. An example of this dependency after model calculations of TAKAKURA (1961) and others is given in fig. 17. Hence a relation between the transitional frequency and the heliographic longitude of the emitting region follows, which, so far as the author is aware, in the average is not yet well established. In general, within a given frequency interval, a reversal of the sense of polarization is only observed within a certain range of directions. After estimations basing on a dipol model of the magnetic field (TAKAKURA, 1961)

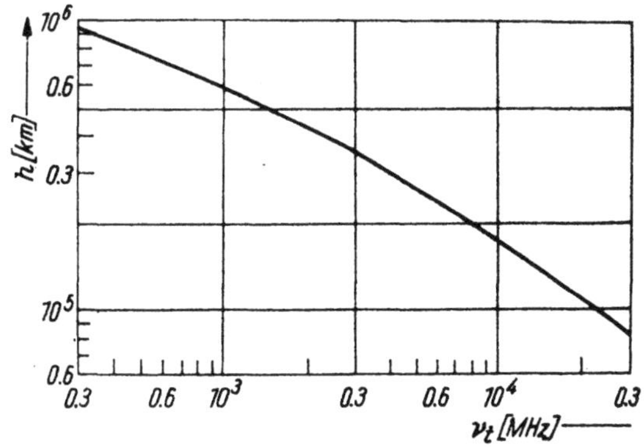

Fig. 17. Dependence of the transitional frequency ν_t of a QT-region on the height h above the photosphere after model calculations of COHEN and DWARKIN (1961) and TAKAKURA (1961) for a subphotospheric dipol ($a=0.5\ R_\odot$), maximal field strength at the photosphere $H_0 = 2000$ gauss) and the BAUMBACH-ALLEN corona

only one fourth of all cases should exhibit such reversal of polarization in the range 1—10 GHz which is less than observed (TANAKA and KAKINUMA, 1959). Besides the propagation hypothesis also explanations basing on processes inherent to the source itself have been propsed. Reversals could be due to a superposition of different sources with different emission characteristics (TANAKA and ENOME, 1970) or to special models of gyro-synchrotron absorption (HOLT and RAMANATY, 1969).

3.3.4. *Absorption under the influence of external magnetic fields*

As mentioned in 3.2.2 the dispersion relation can be used for a calculation of the absorption coefficient. The separation of the refractive index into the real and imaginary part

$$n_j = \mu_j - i\,\chi_j \qquad (3.3.4.\text{-}1)$$

yields the absorption index χ_j, from which the absorption coefficient follows

$$\varkappa_{\nu,j} = \frac{2\pi\nu}{c}\,\chi_j\,. \qquad (3.3.4.\text{-}2)$$

The designation of $\frac{\omega \chi_j}{c}$ as absorption coefficient follows from a consideration of the equation of oscillation for an electron in the following form

$$E_x = E_{x,0} \exp\left\{i\,\omega\left(t - \frac{n_j z}{c}\right)\right\} = \\ = E_{x,0} \exp\left\{-\frac{\omega \chi_j z}{c} + i\,\omega\left(t - \frac{\mu_j z}{c}\right)\right\} \quad (3.3.4.\text{-}3)$$

where the term $\exp\left(-\frac{\omega \chi_j z}{c}\right)$ represents the spatial absorption.

Comparing the results of different authors, it is noticed that because $I_\nu \sim E^2$ sometimes also the twofold value of (3.3.4.-2) is defined as absorption coefficient.

Further, if a remarkable anisotropy is to be taken into account, equation (3.3.4.-2) is to be replaced by

$$\varkappa_{\nu,j} = \frac{\omega}{c}\, \chi_j \cos \vartheta_j\,. \quad (3.3.4.\text{-}4)$$

Now we calculate $\varkappa_{\nu,j}$ from (3.3.4.-2) for the case of wave propagation parallel to the direction of the magnetic field. Starting from

$$n_j^2 = \frac{(1 - X \mp Y_L) - i\,Z}{(1 \mp Y_L) - i\,Z} \quad (3.3.4.\text{-}5)$$

it follows

$$\varkappa_{\nu,j} = \frac{\omega}{c}\left\{\frac{[(1 - X \mp 2\,Y_L \pm X\,Y_L + Y_L^2 + Z^2)^2 + X^2 Z^2]^{1/2} - (1 - X \mp 2\,Y_L \pm X\,Y_L + Y_L^2 + Z^2)}{2\,[(1 \mp Y_L)^2 + Z^2]}\right\}^{1/2}. \quad (3.3.4.\text{-}6)$$

From this for $Z \ll 1$ the well known special case

$$\varkappa_{\nu,j} = \frac{\omega}{2\,c}\,\frac{X\,Z}{(1 \mp Y_L)^2} \quad (3.3.4.\text{-}7)$$

is obtained

Similar for a propagation perpendicular to the external magnetic field we get the expressions

$$\left.\begin{aligned} n_1^2 &= \frac{[(1 - X)^2 - Z^2 - Y_T^2] - i \cdot 2\,Z\,(1 - X)}{[1 - X - Z^2 - Y_T^2] - i\,(2\,Z - X\,Z)} \\ n_2^2 &= \frac{(1 - X) - i\,Z}{1 - i\,Z} \end{aligned}\right\} \quad (3.3.4.\text{-}8)$$

and hence

$$\begin{aligned} \varkappa_{\nu,1} = \frac{\omega}{c}\{(&[[(1 - X - Z^2 - Y_T^2)^2 - (X - X^2)\,(1 - X - Z^2 - Y_T^2) + \\ &+ 2\,(X\,Z - Z)^2 - 2\,Z^2\,(X - 1)]^2 + [(2\,X\,Z - 2\,Z) \times \\ &\times (1 - X - Z^2 - Y_T^2) - (X\,Z - 2\,Z)\,(1 - 2\,X + X^2 - Z^2 - Y_T^2)]^{1/2} - \\ &- [(1 - X - Z^2 - Y_T^2)^2 - (X - X^2)\,(1 - X - Z^2 - Y_T^2) + \\ &+ 2\,(X\,Z - Z)^2 - 2\,Z^2\,(X - 1)]\}^{1/2} \times \\ &\times \{2\,[(1 - X - Z^2 - Y_T^2)^2 + (X\,Z - 2\,Z)^2]\}^{-1/2}\,. \end{aligned} \quad (3.3.4.\text{-}9)$$

The expression obtained for $\varkappa_{\nu,2}$ is identical with (3.2.2.-4) for the absence of magnetic fields.

For $X \ll 1$, $Z \ll 1$ again the often cited approximations follow

$$\left.\begin{aligned} \varkappa_{\nu,1} &= \frac{\omega}{2c} \frac{XZ(1+Y_T^2)}{(1-X-Y_T^2)^2} \\ \varkappa_{\nu,2} &= \frac{\omega}{2c} \frac{XZ}{(1-X)^{1/2}}\,. \end{aligned}\right\} \qquad (3.3.4.\text{-}10)$$

3.3.5. *Optical depth and polarization characteristics of radio bursts*

Now we consider some general consequences of the formulae of the previous section. At first, for simplicity, an isothermal plasma model may be assumed which emits bremsstrahlung under the conditions of the QL-approximation. Then the optical depth is given by

$$\tau_{\nu,j} = \int\limits_{(S)} \frac{\omega_p^2 \nu_{\text{coll}}}{\omega^2 c \left(1 \mp \frac{\omega_H}{\omega} \cos\theta\right)^2} ds \approx \frac{\omega_p^2 \nu_{\text{coll}} \cdot L}{\omega^2 c \left(1 \mp \frac{\omega_H}{\omega} \cos\theta\right)^2} = \frac{\lambda^2}{\lambda_{\text{crit}}^2} \cdot \frac{1}{\left(1 \mp \frac{\omega_H}{\omega} \cos\theta\right)^2} \qquad (3.3.5.\text{-}1)$$

where

$$\lambda_{\text{crit}} = \left\{\frac{4\pi^2 c^3}{\omega_p^2 \nu_{\text{coll}} \cdot L}\right\}^{1/2} \qquad (3.3.5.\text{-}2)$$

(L is a characteristical length of the emitting volume).

The quantity λ_{crit} introduced by ZHELEZNYAKOV (1964) has he following meaning: $\lambda = \lambda_{\text{crit}}$ denotes the wavelength where without magnetic fields $\tau_\nu = 1$.

For $|(\omega_H/\omega)\cos\theta| < 1$ from (3.3.5.-1) it follows immediately

$$\tau_{\nu,j} = \left(\frac{\lambda}{\lambda_{\text{crit}}}\right)^2 \left[1 \pm 2\frac{\omega_H}{\omega}\cos\theta + 3\left(\frac{\omega_H}{\omega}\cos\theta\right)^2\right]. \qquad (3.3.5.\text{-}3)$$

Thus the existence of an external magnetic field takes an effect on the intensity of the emitted bremsstrahlung by a modification of the optical depth. In comparison to the magnetic field free case in particular the longitudinal magnetic field leads to an increase of the intensity of the extraordinary wave mode and to a decrease of the ordinary wave mode, respectively.

If one considers the case of very small optical depths, i.e. $\lambda < \lambda_{\text{crit}}$, after (3.2.3.-11) the intensity of the radiation coming from an emitting region of the optical depth $\tau_{\nu,j}$ is given by

$$\begin{aligned} I_\nu &= I_{\nu,1} + I_{\nu,2} = I_{\nu,0}\left(\tau_{\nu,1} + \tau_{\nu,2} - \frac{\tau_{\nu,1}^2}{2} - \frac{\tau_{\nu,2}^2}{2}\right) \\ &= 2 I_{\nu,0}\left(\frac{\lambda}{\lambda_{\text{crit}}}\right)^2 \left[1 - \frac{1}{2}\left(\frac{\lambda}{\lambda_{\text{crit}}}\right)^2 + 3\left(\frac{\omega_H}{\omega}\cos\theta\right)^2\right] \end{aligned} \qquad (3.3.5.\text{-}4)$$

If

$$\frac{1}{2}\left(\frac{\lambda}{\lambda_{\text{crit}}}\right)^2 < 3\left(\frac{\omega_H}{\omega}\cos\theta\right)^2$$

i.e.

$$\frac{T^{3/2}\,\nu_H^2\cos^2\theta}{\overline{N}_e^2\cdot L} > 3\cdot 10^2 \qquad [\text{cgs}]\,, \qquad (3.3.5.\text{-}5)$$

the spectral distribution of the intensity $(I_{\nu,1} + I_{\nu,2})$ rises with increasing wavelength. Then, in a certain region the spectrum comes to a maximum. The estimation of the position of this maximum as given e.g. by ZHELEZNYAKOV (1964) leads to errors because of the approximation $Y_L < 1, \tau_{\nu,j} \ll 1$ made here. Instead of this, from (3.3.4.-7) a singularity of $\tau_{\nu,1}$ at the gyrofrequency follows and hence the position of the spectral maximum near this point must be expected, which is slightly modified, if the influence of the collision frequency is taken into account. Concerning the amplitude of this maximum the general statement is valid, that at each frequency the intensity cannot exceed the value given by the KIRCHHOFF-PLANCK function.

Without the assumption $\tau_{\nu,j} \ll 1$ instead of (3.3.5.-4) it is obtained

$$I_\nu = 2\,I_{\nu,0}\left\{1 - \exp\left(-\,\varepsilon\,L\,\lambda^2\left[1 + 3\left(\frac{\omega_H}{\omega}\cos\theta\right)^2\right]\right)\cosh\left[2\,\varepsilon\,L\,\lambda^2\frac{\omega_H}{\omega}\cos\theta\right]\right\} \qquad (3.3.5.\text{-}6)$$

with

$$\varepsilon = \frac{\omega_p^2\,\nu_{\text{coll}}}{4\,\pi^2\,c^3} = \frac{\xi\,N_e^2}{c^2\,T^{3/2}}$$

$$\frac{\xi}{c^2} \approx 1.11\cdot 10^{-22} \qquad (\text{cgs}). \qquad (3.3.5.\text{-}7)$$

For $\cos\theta = 1$ it follows

$$I_\nu = 2\,I_{\nu,0}\left\{1 - \exp\left(-\,\varepsilon\,L\,\lambda^2\,(1 + A\,H_L^2\,\lambda^2\right]\cosh\left(2\,\varepsilon\,L\,\lambda^2\frac{\omega_H}{\omega}\right)\right\}, \qquad (3.3.5.\text{-}8)$$

where $A = 2.6\cdot 10^{-8}$ cgs if H is measured in gauss (MOLCHANOV, 1962).

If $H < 3000$ gauss, for short wavelengths (cm-region) the expression (3.3.5.-8) can be simplified as follows:

$$S_\nu \approx 2\,k\,\varepsilon\,L\,T_e\,\Omega\left[1 + \left(A\,H_L^2 - \frac{1}{2}\,\varepsilon\,L\right)\lambda^2\right], \qquad (3.3.5.\text{-}9)$$

where Ω denotes the solid angle of the emitting region seen at the earth. In equations (3.3.5.-8) and (3.3.5.-9) the longitudinal magnetic field H_L and the emission measure $N_e^2\,L$ are unknown quantities. For an estimation of these parameters at least two independent measurements are necessary. If it is attempted to obtain these informations from observations at two neighbouring wavelengths λ_a and λ_b, it can be written

$$A\,H_L^2 - \frac{1}{2}\,\varepsilon\,L = \frac{S_{\nu_a} - S_{\nu_b}}{S_{\nu_b}\,\lambda_a^2 - S_{\nu_a}\,\lambda_b^2} \qquad (3.3.5.\text{-}10)$$

and

$$\varepsilon L = \frac{S_{\nu_b} \lambda_a^2 - S_{\nu_a} \lambda_b^2}{(\lambda_a^2 - \lambda_b^2)\, 2\, k\, \Omega\, T_e} \tag{3.3.5.-11}$$

and hence

$$N_e^2 L = \frac{(S_{\nu_b} \lambda_a^2 - S_{\nu_a} \lambda_b^2)\, c^2\, T_e^{1/2}}{(\lambda_a^2 - \lambda_b^2)\, 2\, k\, \Omega\, \xi} . \tag{3.3.5.-12}$$

If $S_\nu'(\lambda) = \frac{\partial S_\nu}{\partial \lambda}$ denotes the gradient of the spectrum at a given wavelength λ, one obtains

$$\left(A\, H_L^2 - \frac{1}{2}\, \varepsilon\, L\right) = \frac{S_\nu'}{4\, k\, \Omega\, \lambda\, T_e\, \varepsilon\, L} . \tag{3.3.5.-13}$$

If a noticeable gradient of the spectrum exists, i.e. if $A\, H_L^2 > \frac{1}{2}\, \varepsilon\, L$, the magnetic field is given by

$$H_L = \frac{1}{\left[2\, A\, \lambda \left(\frac{S_\nu}{S_\nu'} - \frac{\lambda}{2}\right)\right]^{1/2}} . \tag{3.3.5.-14}$$

Analogeously also formulae for a transversal magnetic field H_T could be derived.

The above given expressions have been used by MOLCHANOV (1961) for an investigation of the S-component of solar radio emission. But later from the improved knowledge of the polarization properties of the radio emission it became apparent, that the spectral maximum of the S-component near $\lambda \lesssim 10$ cm cannot be explained by bremsstrahlung. Nevertheless, for special cases of gradual cm-bursts possibly a certain amount of bremsstrahlung has to be considered.

Now some remarks on the polarization characteristics will be given. Considering circular polarization, for a homogeneous and isothermal burst model from (3.3.2.-9) it follows

$$\varrho_c = \frac{S_{\nu,1} - S_{\nu,2}}{S_{\nu,1} + S_{\nu,2}} = \frac{T_{e,1} - T_{e,2}}{T_{e,1} + T_{e,2}} = \frac{e^{-\tau_{\nu,2}} - e^{-\tau_{\nu,1}}}{2 - e^{-\tau_{\nu,1}} - e^{-\tau_{\nu,2}}} = \frac{\sinh \Delta\tau_\nu}{e^{\bar{\tau}_\nu} - \cosh \Delta\tau_\nu} \tag{3.3.5.-15}$$

with

$$\bar{\tau}_\nu = \frac{\tau_{\nu,1} + \tau_{\nu,2}}{2}, \qquad \Delta\tau_\nu = \frac{\tau_{\nu,1} - \tau_{\nu,2}}{2} . \tag{3.3.5.-16}$$

Using (3.3.4.-7) for a constant longitudinal magnetic field and $Z \ll 1$ the following relations are obtained

$$\tau_{\nu,1} = \frac{\tau_\nu^0}{(1 - Y_L)^2}, \qquad \tau_{\nu,2} = \frac{\tau_\nu^0}{(1 + Y_L)^2} \tag{3.3.5.-17}$$

and

$$\bar{\tau}_\nu = \tau_\nu^0 \frac{1 + Y_L}{(1 - Y_L)^2}, \qquad \Delta\tau_\nu = \tau_\nu^0 \frac{2\, Y_L}{(1 - Y_L)^2}, \tag{3.3.5.-18}$$

where τ_ν^0 denotes the optical depth in the magnetic field free case.

For $\tau_\nu^0 \ll 1$ the well known relation follows

$$\varrho_c = \frac{2\,Y_L}{(1 + Y_L^2)}\,. \tag{3.3.5.-19}$$

If $Y_L \ll 1$, i.e. at greater distances from the gyrofrequency it becomes

$$\varrho_c \sim \lambda\,. \tag{3.3.5.-20}$$

On the other side, for $\tau_\nu^0 \gg 1$ one obtains

$$\varrho_c = 0\,. \tag{3.3.5.-21}$$

Thus, under the above made assumptions the influence of the magnetic field results in a maximum of the degree of circular polarization lying between the cases (3.3.5.-20) and (3.3.5-21). For large optical depths (black body radiation) no appreciable polarization can be expected.

From the above given relations within the range of assumed conditions the magnetic field can be estimated from polarization measurements at different wavelengths. Certain methods in this direction have been treated by GELFREIKH (1962a), KOROLKOV et al. (1960) and others. But it appears difficult to take into account both, the generation of the polarized radiation and the influence of the propagation through the solar atmosphere.

Other proposals have been made with respect to an application of the degree of eliptical polarization for a calculation of the magnetic field strength in the burst sources (GELFREIKH, 1962b). All these methods are restricted to the case of weak coupling of the electromagnetic wave modes.

At least, analogeously to (3.3.5.-15) for a transversal magnetic field the degree of polarization is given by

$$\varrho_l = \frac{3\,Y_T^2}{2 - Y_T^2}\,. \tag{3.3.5.-22}$$

3.4. Gyroemission

3.4.1. Gyroemission as special case of magnetic bremsstrahlung (cyclotron radiation)

After the consideration of the bremsstrahlung now the second large group of emission mechanisms, the gyro-synchrotron or cyclotron radiation is to be discussed. In this case no longer the COULOMB potential of positive ions or protons but the LORENTZ force has the main influence on the motion of the charged particles which emit radiation. Instead of cyclotron radiation also sometimes the denotation "magnetic bremsstrahlung" is used which is illustrative but should not lead to confusion with the pecularities of COULOMB-bremsstrahlung in magnetic fields.

For non-relativistic electron energies $\beta \ll 1$ the classical theory (cf. e.g. LANDAU and LIFSHITZ, 1961; IVANENKO and SOKOLOV, 1953; SOKOLOV and TERNOV, 1966) delivers a dipol radiation at the electron gyrofrequency

$$\nu_H = \frac{e\,H_T}{2\,\pi\,m_e\,c}\,, \tag{3.4.1.-1}$$

which in this approximation depends on the magnetic field strength only. The radius of the gyration is obtained from

$$\frac{2\,\pi\,r_H}{\lambda_H} = \beta\,, \qquad \lambda_H = \frac{c}{\nu_H}\,, \qquad \beta = \frac{v_e}{c}\,. \tag{3.4.1.-2}$$

It is expected, that for solar applications the emission at the fundamental frequency ν_H is of no importance, because the extraordinary wave cannot propagate at the gyrofrequency as it follows from the dispersion relation. Therefore, the effect of the emission of harmonics of the gyrofrequency is of greater importance.

As it is well known from the MAXWELL-LORENTZ theory, by D'ALEMBERT's solution of the wave equation

$$\square\, A_\mu \equiv \left(\nabla^2 - \frac{1}{c^2}\frac{\partial^2}{\partial t^2}\right) A = -\frac{4\,\pi\,e}{c} \int P(s)\,(x - \xi_\mu)\,\dot{\xi}_\mu(s)\,ds \tag{3.4.1.-3}$$

for a point-like electron the following expressions for the retarded potentials are obtained

$$\left.\begin{aligned} \frac{A_4}{i\,c} &= e \int \frac{\delta\left(\tau - t + \frac{r(\tau)}{c}\right)}{r(\tau)}\,d\tau \\ A_\varkappa &= \frac{e}{c} \int v(\tau)\,\frac{\delta\left(\tau - t + \frac{c(\tau)}{c}\right)}{r(\tau)}\,d\tau \\ \chi &= 1,\ 2,\ 3 \end{aligned}\right\} \tag{3.4.1.-4}$$

and the integration over the time τ give the well known LIÉNARD-WIECHERT potentials. Here the different retardations observed at a fixed point $\boldsymbol{r}$ with respect to different points of the orbit of the electron are of special interest which explains the occurrence of harmonics for sufficient high velocities $\boldsymbol{v}$. This effect is described by the delta function in (3.4.1.-4) which can be written in the form

$$\delta(t') = \frac{\omega}{2\,\pi} \sum_{s=-\infty}^{\infty} e^{-i\,s\,\omega\,t}\,. \tag{3.4.1.-5}$$

Accordingly the potentials, in particular the vector potential of (3.4.1.-4), can be respresented by the sum $\sum A_s \exp(-\,i\,s\,\varphi_s)$ where the coefficients A_s and the phases φ_s finally lead to the estimation of the emission propabilities.

Furthermore it must be stated that for $\beta \gg 0$ as a consequence of the relativistic DOPPLER effect the observed frequencies ν_s are dependent not only on the strength of the magnetic field but essentially become also a function of the particle velocity or energy. This is described by the so-called emission equation

$$\nu_s(\beta, \theta, \Phi) = \frac{s\,\nu_H\,(1-\beta^2)^{1/2}}{1 - n_j\,\beta\cos\theta\cos\Phi}\,. \tag{3.4.1.-6}$$

In this equation ν_H denotes the non-relativistic gyrofrequency given by (3.4.1.-1). The angles θ and Φ are defined as follows

$$\left.\begin{aligned} \theta &= \sphericalangle\,(\boldsymbol{k}_j, \boldsymbol{H}) \\ \Phi &= \sphericalangle\,(\boldsymbol{v}_e, \boldsymbol{H})\,. \end{aligned}\right\} \tag{3.4.1.-7}$$

A graphical representation of the emission equation is given in fig. 18., where for the limiting cases $\Phi = 0$ and $\Phi = \frac{\pi}{2}$, $n_j = 1$ in a three-dimensional diagram the quantity $\frac{\nu_s}{H}$ is plotted against β and θ.

In many applications the refractive index n_j contained in (3.4.1.-6) is assumed to be unity. But there are also important cases for which the influence of the medium expressed by $n_j \neq 1$ must be taken into account (cf. 3.5.3).

For

$$s > 0\,, \qquad n_j\,\beta\cos\theta\cos\Phi < 1 \tag{3.4.1.-8}$$

the so-called normal DOPPLER effect is realized, where $v\cos\Phi < v_{k_j}$.

The other case

$$s < 0\,, \qquad n_j\,\beta\cos\theta\cos\Phi > 1 \tag{3.4.1.-9}$$

corresponds to the anomalous DOPPLER effect, which takes place if $v\cos\Phi > v_{k_j}$ in a medium with $n_j > 1$ is fulfilled.

For the normal DOPPLER effect the radiation corresponding to $\nu_s > s \times \nu_H\,(1-\beta^2)^{1/2}$ or $\nu_s < s\,\nu_H\,(1-\beta^2)^{1/2}$ is propagated into or against the hemisphere of the direction of the particle motion ($\beta\cos\theta > 0$ or $\beta\cos\theta < 0$), respectively. Both cases simultaneously can be realized. On the other side, for the anomalous DOPPLER effect only forward emission ($\beta\cos\theta > 0$) is possible.

The case $s = 0$ describes the ČERENCOV radiation, for which the condition $\cos\theta = 1/(n_j\,\beta\cos\Phi)$ is valid. Here the radiation is emitted into the hemisphere of the direction of the particle motion as the anomalous DOPPLER effect.

Let us now consider at first the case $s > 0$, which is of a great interest for radio astronomy. According to EIDMANN (1958) the energy quantum emitted by one electron per time unit into a solid angle element $d\Omega$ is given by

$$dW_s = \sum_{j=1,2} \frac{(n_j\,e^2\,\omega^2\,\gamma_j^2\,\{\beta\sin\Phi\,I_s'(s\,\alpha) + [a_j\,\beta\sin\Phi\,\alpha^{-1} + b_j\,\beta\cos\Phi]\,I_s\,(s\,\alpha)\}^2}{2\,\pi\,c\left|1 - \beta\cos\Phi\cos\theta\left(n_j + \omega\dfrac{d\,n_j}{d\,\omega}\right)\right|} \tag{3.4.1.-10}$$

$$dW = \sum_{s=1}^{\infty} dW_s = \boldsymbol{S}\,d\Omega \tag{3.4.1.-11}$$

($\boldsymbol{S}$ — POYNTING vector),

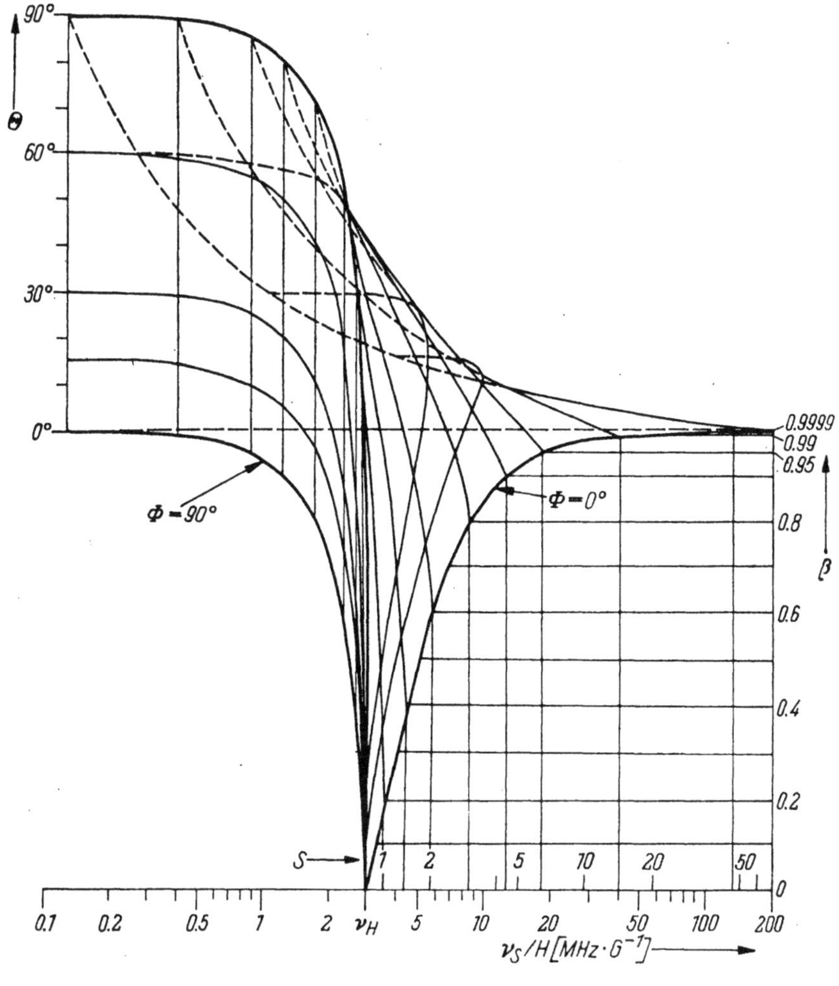

Fig. 18. Three-dimensional representation of the emission equation of the Doppler-shifted gyrofrequencies

where

$$\left.\begin{aligned} \gamma_j^2 &= \frac{1}{1 + K_j^2} \\ \alpha &= \frac{n_j \beta \sin \Phi \sin \theta}{1 - n_j \beta \cos \Phi \cos \theta} \end{aligned}\right\} \qquad (3.4.1.\text{-}12)$$

and

$$\left.\begin{aligned} a_j &= K_j \cos \theta + \Gamma_j \sin \theta \\ b_j &= \Gamma_j \cos \theta - K_j \sin \theta \,, \end{aligned}\right\} \qquad (3.4.1.\text{-}13)$$

K_j and Γ_j are explained by equations (3.3.2.-14) and (3.3.2.-19). The index $j = 1$ again denotes the extraordinary and the index $j = 2$ denotes the ordinary wave component, respectively.

For n_j the dispersion relations given in section 3.3.1 are valid. $I_s(s\,\alpha)$ and $I_s'(s\,\alpha)$ are BESSEL functions and their derivatives with respect to the argument s, respectively. It should be mentioned that in (3.4.1.-10) the thermal motion of the electrons was neglected.

For $Y \ll 1$, i.e.

$$s \gg \frac{1 - n_j\,\beta \cos \Phi \cos \theta}{(1 - \beta^2)^{1/2}} \tag{3.4.1.-14}$$

there follows from (3.4.1.-10)

$$dW_s = \frac{n\,e^2\,\omega^2 \left\{\beta^2 \sin^2 \Phi\,[I_s'(s\,\alpha)]^2 + \left(\frac{\cos\theta - \beta\cos\Phi}{n\,\beta \sin\theta \sin\Phi}\right)^2 [I_s(as\,\alpha)]^2\right\}}{2\,\pi\,c \left|1 - \beta \cos\left(\Phi \cos\theta\left(n + \frac{1}{n\,\omega^2}\right)\right)\right|}\,d\Omega \tag{3.4.1.-15}$$

where now $n = n_1 = n_2$ is assumed.

Under the assumption $n = 1$ it follows

$$dW_s = \frac{e^2\,\omega_H^2\,(1 - \beta^2)\,s^2\,(\beta \sin\Phi)^2 \left\{[I_s'(s\,d)\,]^2 + \frac{\cos\theta - \beta\cos\Phi}{\beta \sin\theta \sin\Phi}\,[I_s(s\,\alpha)]^2\right\}}{2\,\pi\,c\,|1 - \beta\cos\Phi\cos\theta|}\,d\Omega \tag{3.4.1.-16}$$

(ZHELEZNYAKOV, 1964). This result is substantially in accordance with a corresponding formula obtained by TAKAKURA (1961) from a generalization of SCHWINGER's (1949) calculations, but differs from that by a factor $|1-\beta\cos\Phi \times \times \cos\theta|^{-1} \approx \sin^{-2}\Theta$ (cf. (3.4.1.-27)).

For $\Phi = \frac{\pi}{2}$, i.e. electron motion in circular orbits, the well known expression follows

$$dW_s = \frac{e^2\,s^2\,\beta^2\,c}{2\,\pi\,r_H}\,\{\beta^2[I_s'(s\,\beta \sin\theta)]^2 + \mathrm{ctg}^2\,\theta[I_s(s\,\beta\sin\theta)]^2\}\,d\Omega. \tag{3.4.1.-17}$$

For an estimation of the emission coefficient η_ν the knowledge of the energy distribution $N_s(\beta)$ of the electrons is needed contributing to the emission on the sth harmonic number at a certain frequency ν. Then we have

$$\eta_\nu = \sum_s N_s(\beta)\,dW_s\,, \tag{3.4.1.-18}$$

where $\eta_\nu = \eta_\nu(\beta, \theta, \Phi, H)$. In general numerical computations are necessary in order to derive η_ν from known or assumed $N_s(\beta)$. For an estimation of the energy spectrum

$$W(\omega) = \sum_{s=1}^{\infty} \delta\,(\omega - s\,\omega_H)\,W_s \quad [\mathrm{erg\ s^{-1}}] \tag{3.4.1.-19}$$

the integral

$$W_s(\beta) = \int dW_s = \int dS_{\nu,s}\,d\Omega \tag{3.4.1.-20}$$

must be solved. Using (3.4.1.-17) the result can be written as

$$W_s(\beta) = \frac{e^2 \beta c s}{r_H^2} \left\{ 2 \beta^2 I'_{2s}(2 s \beta) - (1 - \beta^2) \int_0^{2s\beta} I_{2s}(x) \, dx \right\}, \quad (3.4.1.\text{-}21)$$

which is the famous formula already given by SCHOTT in 1912.

However, EIDMAN's results as given above were completed for relativistic energies and slightly corrected by LIEHMON (1965). He obtained the following expressions

$$d W_s = \frac{e^2}{2 \pi c} \cdot \frac{\omega^2 n_j^2 K^2 \left[- \beta_T I'_s(x_s) + \frac{a_y s \beta_T}{x_s + a_z \beta_L} I_s(x_s) \right]^2 \left(1 + \frac{\omega}{n_j} \frac{d n_j}{d \omega} \right)}{\left| 1 - \beta_L n_j \cos\theta \left(1 + \frac{\omega}{n_j} \frac{\partial n_j}{\partial \omega} \right) \right|} \quad (3.4.1.\text{-}22)$$

where $K = \frac{1}{\sqrt{1 + a_\theta^2}}$ $\left(\text{instead of } \frac{1}{2} \text{ by EIDMAN (1958)}\right)$ and (3.4.1.-23)

$$\left.\begin{aligned} a_y &= a_\theta \cos\theta + a_K \sin\theta \\ a_z &= a_K \cos\theta - a_\theta \sin\theta \\ a_\theta &= \frac{- y_s \cos\theta}{y_s^2 + \frac{(\omega_p/\omega_H)^2}{n_j^2 - 1}} \\ a_K &= \frac{- y_s \sin\theta \, (n_j^2 - 1)}{\frac{\omega_p^2}{\omega_H^2} - y_s^2} \end{aligned}\right\} \quad (3.4.1.\text{-}24)$$

$$y_s = \frac{s \sqrt{1 - \beta^2}}{1 - n_j \beta_L \cos\theta}, \qquad x_s = \frac{\beta_T n_j y_s \sin\theta}{\sqrt{1 - \beta^2}} \quad (3.4.1.\text{-}25)$$

(cf. also YIP, 1967).

The expression (3.4.1.-22) differs from (3.4.1.-10) by the special form of the factors in the numerator and the sign on $I'_s(x_s)$ as well as by the term taking into account the frequency dependence of n_j in the denominator.

If $\beta_L = 0$, i.e. $\Phi = \frac{\pi}{2}$, it is obtained

$$d W_s = \frac{e^2}{2 \pi c} \omega^2 n_j K^2 \left[- \beta_T I'_s(x_s) + a_y s \beta_T \frac{I_s(x_s)}{x_s} \right]^2 \left(1 + \frac{\omega}{n_j} \frac{\partial n_j}{\partial \omega} \right) \quad (3.4.1.\text{-}26)$$

From this in vacuum with $n = 1$, $a_\theta = \pm 1$, $a_K = 0$ again SCHOTT's formula results.

Concerning the calculation of the radiation at arbitrary pitch angles $\Phi \neq \frac{\pi}{2}$ it remains somewhat insufficient that an exact description of the motion of an electron in a spiral (helical) orbit is not yet developed, though corresponding equations to those of a circular motion can be obtained by a LORENTZ transform as noted by SOKOLOV et al. (1968).

If quantum effects are taken into account, effects like spin polarization occur and generalized LAGUERRE polynoms appear instead of the BESSEL functions, which is not considered in the classical approximation where $h\nu \ll 1$. With regard to the quantum mechanical treatment SOKOLOV et al. (1968) critized the results of GINSBURG and SYROVATSKY (1965) and WESTFOLD (1959) and proposed a correction factor $(1 - \beta \cos \Phi \cos \theta)^{-1}$ for the emitted frequency of cyclotron radiation, but this factor is not restricted to quantum physics and already contained in the emission equation (3.4.1.-6).

The correction factor of $\sin^{-2} \theta$ for the synchrotron radiation received from highly relativistic electrons has been suggested by EPSTEIN and FELDMAN (1967), which was based on the relation

$$dW_{\text{received}} = \frac{dW_{\text{emitted}} \cdot t}{t \cdot \frac{c\,t \cdot \cos^2 \theta}{c}} = \frac{dW_{\text{emitted}}}{\sin^2 \theta} \qquad (3.4.1.\text{-}27)$$

(cf. fig. 19). Nevertheless, it was pointed out by SCHEUER (1968) and TAKAKURA and UCHIDA (1968) that there is no need for a revision of the formulae of

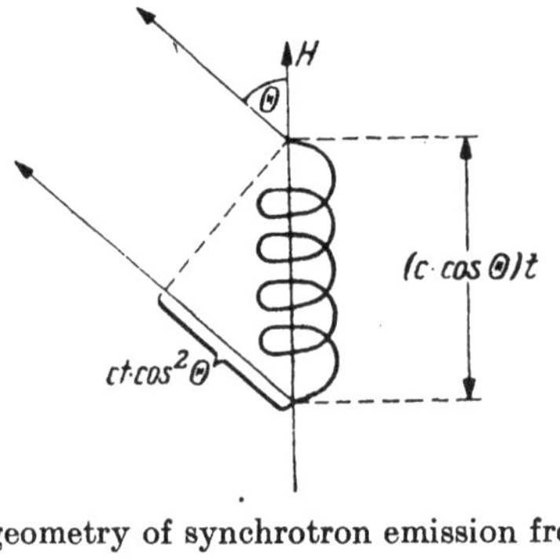

Fig. 19. On the geometry of synchrotron emission from helical orbits

dW_{emitted} of the cyclotron radiation as well as for a "re-interpretation" of the results derived with the help of these formulae. As a matter of fact, dW_{received} can be computed in two ways: Either by multiplying dW_{emitted} according to the conventional formulae with the mean numbers of electrons in the source taking into account an appropriate angle θ, or by multiplying the EPSTEIN-FELDMAN formulae with the mean number of electrons contributing to the radiation observed at a particular instant.

3.4.2. *Gyro-resonance absorption*

The formulae given in the foregoing section are somewhat awkward for a practical application because the number s occurs as well in the order of the BESSEL functions as in their argument. Therefore often approximations for low and high particle energies are used being more convenient for numerical computations.

Here we start with the case of low electron energies. The high energy approximation leading to the synchrotron emission formulae will be discussed in the next section.

As mentioned in sections 3.2.2 and 3.3.4, the calculation of the absorption and emission coefficients can be achieved in several ways. One method consists in using the emission formulae of the one particle approximation as given in the previous section and calculating the absorption coefficient by the assumption of thermal equilibrium from KIRCHHOFF's law. Then the energy distribution is identical with the MAXWELL distribution over which is integrated. This treatment corresponds to the hydrodynamic approximation. The advantages of this method are in the relatively simple mathematical treatment. But it appears as a serious disadvantage that for the neglections usually adopted the approximations of the formulae for the emission at the fundamental gyrofrequency are imperfect.

A more reliable but also more complicated method is based on the kinetic theory. Here from the (mostly linearized) BOLTZMANN (or VLASOV) or FOKKER-PLANCK equation in combination with the MAXWELL field equations dispersion relations are derived (cf. 3.3.1) from which the complex refraction index and hence the absorption coefficient follow.

At first we consider the main steps of the first method, which practically is equivalent to the method of the EINSTEIN coefficients (3.2.2.-6) leading in the case of thermal equilibrium to KIRCHHOFF's law.

Corresponding to the assumption $\beta \ll 1$ and the following neglections

$$\left.\begin{array}{c} n_j\,\beta\cos\Phi \ll 1 \\ |l| \equiv |s\,n_j\,\beta\sin\Phi\sin\theta| \ll 1 \end{array}\right\} \tag{3.4.2.-1}$$

i.e. restriction to the first term of the expansion of the BESSEL functions $I_s(s\,\alpha)$ and $I_s'(s\,\alpha)$ from (3.4.1.-10) it follows for the dipol emission

$$d W_s = \frac{n_j\,e^2\,\omega^2\,\gamma_j^2\,[s\,\beta\sin\Phi\,(1+a_j) + l\,b_j\,\beta\cos\Phi]^2\,e^{2s-2}}{2\pi c\left|1 - \beta\cos\Phi\cos\theta\left(n_j + \dfrac{\omega\,\partial n_j}{\partial\omega}\right)\right|2^{2s}(s!)^2} \tag{3.4.2.-2}$$

$$d W_s \approx \frac{e^2\,\omega_H^2\,\beta^2\sin^2\theta}{4\pi c}\,[n_j\,\gamma_j^2\,(1+a_j)^2\,(s\,n_j\,\beta\sin\Phi\sin\theta)^{2s-2}]_{\omega=s\,\omega_H}\,2^{-2s+1}\,s^6\,(s!)^{-2} \tag{3.4.2.-3}$$

In this equation the approximation

$$(1 + a_j + n_j\,b_j\,\beta\cos\Phi\sin\theta) \approx (1+a_j)_{\omega=s\,\omega_H} \tag{3.4.2.-4}$$

has been used. This is not correct if $s = 1$, because in general we have

$$(1+a_j)_{\omega=\omega_H} = 0\,. \tag{3.4.2.-5}$$

Therefore in the case $s = 1$ annother form of approximation is necessary. ZHELEZNYAKOV (1964) gives the expression

$$dW_{s=1} = \frac{e^2 \omega_H^2 \beta^2 \sin^2 \Phi}{8 \pi c} (1 - Y)^2 \left[n_j \gamma_j \left(\frac{1 + a_j}{1 - Y} + b_j \operatorname{tg} \theta \right)^2 \right]_{\omega = \omega_H} \tag{3.4.2.-6}$$

which is not valid vor $\theta \to \frac{\pi}{2}$ and $n_j \to 0$ as well as for $\theta \to 0$ and $n_j \to \infty$.

It should be noted that the thermal motion of the electrons is not contained in the expressions for n_j^2, γ_j^2, a_j, and b_j. Its consideration gives corrections of the order of β_{th} or β_{th}^2 which with the exception of the case $s = 1$ are neglectible for $\beta_{\text{th}} \ll 1$.

It must be further noted that putting $H \to 0$ ($n_j = 1 - X$) it follows

$$d\dot{W}_s = \frac{n_j e^2 \omega^2 \beta^2 \sin \Phi}{4 \pi c} (1 + \cos^2 \theta)\, 2^{-2s+1} s^4 (s!)^{-2} (s\, n_j \beta \sin \Phi)^{2s-2}. \tag{3.4.2.-7}$$

In the presence of an approximately isotropic medium therefore (since $(1 - Y)^2 \sim \sim \beta^2$), the fundamental frequency is much stronger expressed as in the case of a strong magneto-active plasma. The above mentioned relations show that roughly for

isotropy with	$s \geqq 1$	$dW_s \sim \beta^{2s}$
anisotropy with	$s = 1$	$dW_s \sim \beta^4$
	$s > 1$	$dW_s \sim \beta^{2s}$.

This is in contradiction to a result of TWISS and ROBERTS (1958) who found for the fundamental frequency of the extraordinary wave mode a dependency of the form $dW_1 \sim \beta^2$.

If a system of cylindrical coordinates is used, by integrations over the distribution function $f(v_e)$ and the angle φ the emission coefficient is given by

$$\eta_{\nu,j} = \int_{v_T} \int_{v_L} \int_0^{2\pi} dW_s f(v_e)\, v_T\, dv_L\, dv_T\, d\varphi \tag{3.4.2.-8}$$

with $v_e^2 = v_L^2 + v_T^2$, $v_L = v \cdot \sin \Phi$, $v_T = v \cdot \cos \Phi$.

Using the approximations

$$\beta_{\text{th}}^2 \ll 1\,, \quad \beta^2 \cos^2 \Phi \ll 1\,, \quad |n_j \beta \cos \Phi| \ll 1\,, \quad |s\, n_j \beta_{\text{th}} \sin \theta| \ll 1\,,$$

$$\left| \beta \cos \Phi\, \omega \frac{\partial n_j}{\partial \omega} \right| \ll 1\,, \quad |n_j \cos \theta| \ll |\beta \cos \Phi|$$

there follows

$$\eta_{\nu,j} = \frac{s^{2s} e^2 N_e \omega}{2^s (2\pi)^{5/2} s!\, c^{2s}} \left(\frac{kT}{m_e} \right)^{s - \frac{1}{2}} \frac{\sin^{2s-2} \theta}{|\cos \theta|} n_j^{2s-s} \gamma_j^2$$
$$\times (1 + a_j + n_j b_j \beta \cos \Phi \cos \theta)^2 \exp \left\{ \frac{m_e v_L^2}{2kT} \right\}. \tag{3.4.2.-9}$$

Application of KIRCHHOFF's law leads to the coefficient of gyro-resonance absorption

$$\varkappa_{\nu,js} = B_{js} \frac{s^{2s}\,\omega}{2^s\, s!\, c} \beta_{\text{th}}^{2s-3} \exp\left(\frac{-z_{js}^2}{2}\right) |\cos \vartheta| \,, \qquad (3.4.2.\text{-}10)$$

where

$$\left.\begin{aligned} B_{js} &= \sqrt{\frac{\pi}{2}}\, X\, n_j^{2s-4}\, \gamma_j^2\, [(1 + a_j + b_j\,(1 - s\,Y)\,\text{tg}\,\theta]^2 \frac{\sin^{2s-2}\theta}{\cos\theta} \\ z_{js} &= \frac{\beta\cos\Phi}{\beta_{\text{th}}} = \frac{1 - s\,Y}{n_j\,\beta_{\text{th}}\cos\theta}\,, \qquad \beta_{\text{th}} = \left(\frac{k\,T}{m_e\,c^2}\right)^{1/2}. \end{aligned}\right\} \qquad (3.4.2.\text{-}11)$$

It should be noted that again for $s = 1$ the above given expression must be considered as a rough approximation. In particular outside the range

$$z_{j,1}^2 > 2$$

the errors become large.

Taking $X \ll 1$, i.e. $n_j^2 \approx 1$, $\gamma_j^2 \approx 1/2$, and $\cos\theta = 1$ for $s \geqq 2$ it is obtained

$$\varkappa_{\nu,js} = \frac{\pi}{2} \frac{s^{2s}\,\nu\,X}{2^s\, s!\, c} \beta_{\text{th}}^{2s-3} \frac{(1 \pm \cos\theta)^2 \sin^{2s-2}\theta}{\cos\theta} \exp\left\{-\frac{(1 - s\,Y)^2}{2\,(\beta_{\text{th}}\cos\theta)^2}\right\}. \qquad (3.4.2.\text{-}12)$$

Comparing these formulae with those derived basing on the kinetic theory, e.g. by STEPANOV (1958) it must be noted that in the kinetic derivations often an amplitude attenuation coefficient or an absorption coefficient in direction of k_j are obtained, while in the here given equations the absorption coefficient is taken in direction of the group velocity $d\omega/dk_j$. When these circumstances are taken into account a relatively good agreement of the results of both methods can be stated, so far as the case $s > 1$ ist considered. At the fundamental frequency larger discrepancies are present so that here the treatment of the kinetic theory leads to correct results only.

For comparison, expressions for the first three gyro-resonance absorption coefficients according to GERSHMAN (1960) (cf. also GINSBURG, 1967) obtained from linearized kinetic equations are given below

$$\begin{aligned} \varkappa_{\nu,js=1} &= \sqrt{\frac{2}{\pi}} \frac{k_j\,\beta_{\text{th}} \dfrac{\cos\theta}{X\,n_j}}{2\,X - 2 - \sin^2\theta + 2\,n_j^2 \sin^2\theta} \left\{\left[1 - \left(1 - \frac{7}{4}\sin^2\theta\right)X\right] n_j^4 - \right. \\ &\quad - \left[2 + X\left(\frac{7}{4}\sin^2\theta - \frac{5}{2}\right) + \frac{1}{4}X^2\,(2\cos^2\theta - \text{tg}^2\,\theta)\right] n_j^2 + \\ &\quad \left. + \left[1 - \frac{3}{2}X + \frac{1}{2}X^2\,(1 - \text{tg}^2\,\theta) + \frac{1}{4}X^3\,\text{tg}^2\,\theta\right]\right\}, \\ \varkappa_{\nu,js=2} &= \sqrt{\frac{\pi}{8}} \frac{k_j\,X\,\beta_{\text{th}}\,n_j \sin^2\theta}{\cos\theta} B(Y)_{Y=\frac{1}{2}} \exp\left\{\frac{-(1 - 2\,Y)^2}{2\,\beta_{\text{th}}^2\,n_j^2\cos^2\theta}\right\}_{Y=\frac{1}{2}}, \\ \varkappa_{\nu,js=3} &= \frac{27}{8}\sqrt{\frac{\pi}{8}} \frac{k_j\,X\,\beta_{\text{th}}^3\,n_j^3 \sin^4\theta}{\cos\theta} B(Y)_{Y=\frac{1}{3}} \exp\left\{\frac{-(1 - 3\,Y)^2}{2\,\beta_{\text{th}}^2\,n_j^2\cos^2\theta}\right\}_{Y=\frac{1}{3}}, \end{aligned} \qquad (3.4.2.\text{-}13)$$

where

$$B(Y) = \frac{Y^2 - 1}{Y^2 n_j^2} \left\{ \frac{1}{2} n_j^4 \sin^2 \theta + \left[X \left(\frac{1}{2} + \frac{1}{2} \cos^2 \theta + \frac{\sin^2 \theta}{1 + 4} \right) - \left(1 + \frac{1}{2} \sin^2 \theta \right) \right] n_j^2 \right.$$

$$\left. + (1 - X) \left(1 - \frac{X}{1 + Y} \right) \right\} [2 (1 - Y^2 - X + Y^2 X \cos^2 \theta) n_j^2 - 2 (1 - X)^2$$

$$- (1 + \cos^2 \theta) Y^2 x + 2 Y^2]^{-1}. \qquad (3.4.2.\text{-}14)$$

It should be noted that the temperature dependency of the gyro-resonance emission is proportional to $T^{s-3/2}$ in contrast to the COULOMB bremsstrahlung where $\varkappa_{\nu,j} \sim T^{-3/2}$. From this dependency an increasing influence of the gyro-resonance emission with growing temperature follows. Broad band spectra are generated if spatial distributions of the magnetic field and other plasma parameters in real source regions are considered. Accordingly the mechanism of gyro-resonance absorption has been applied to the interpretation of the S-component of solar radio emission, whose spectral and polarization characteristics by this became explicable (KAKINUMA and SWARUP, 1962; ZHELEZNYAKOV, 1962). Analogeously also for weak energy continuum bursts a portion of gyro-resonance emission is to be expected.

The number of harmonics coming to an effect is determined mainly by the factor β^{2s-3} whereas the exponential factor is restricting the emitted frequency intervals to regions near the harmonics of the gyrofrequency. According to the formulae above given, in the non-relativistic case only a relatively small number of harmonics should occur. This does not fit very well an empirical result of LANDAUER (1962), who detected in a thermal helium-plasma harmonics up to 40 times of the gyrofrequency.

Future efforts should be devoted to a theoretical treatment of non-equilibrium gyroemission, e.g. for a two-peak energy distribution of the radiating electrons which is mathematically more complicated but promises a better approximation of real burst emissions.

3.4.3. Radiative transfer in a magnetoactive plasma

Now we consider briefly the polarization characteristics of cyclotron emission in form of tensor relations. In this notation the polarization is expressed by the tensor

$$I_{rq} = \overline{E_r E_q^*} = \frac{1}{2} \begin{pmatrix} I + Q & U + i V \\ U - i V & I - Q \end{pmatrix} \qquad (3.4.3.\text{-}1)$$

(cf. e.g. LANDAU and LIFSHITZ, 1963). The indices r, q denote the Cartesian coordinates x or y in a system where again the z-axis is in direction of the wave vector k_j.

Accordingly the STOKES parameters are explicitely given by

$$\left.\begin{aligned} I &= \overline{E_x E_x^*} + \overline{E_y E_y^*} = I_{xx} + I_{yy} \,, \\ Q &= \overline{E_x E_x^*} - \overline{E_y E_y^*} = I_{xx} - I_{yy} \,, \\ U &= \overline{E_x E_y^*} + \overline{E_y E_x^*} = I_{xy} + I_{yx} \,, \\ V &= i\,(\overline{E_y E_x^*} - \overline{E_x E_y^*}) = i\,(I_{yx} - I_{xy}) \,, \end{aligned}\right\} \qquad (3.4.3.\text{-}2)$$

where for simplicity $E_z = 0$ and $n = 1$ has been assumed. This restriction can be overcome by changing the definition of I_{rq} and replacing the vector $\boldsymbol{E}$ by the vector of the electric induction $\boldsymbol{D}$, for which $D_z = 0$ also in a magnetoactive plasma is valid (cf. also ZHELEZNYAKOV, 1968).

The equation of radiative transfer can be written

$$\frac{dI_{iq}}{dz} = S_{iq} + K_{iqlm} I_{lm} + R_{iqlm} I_{lm} \,, \qquad (3.4.3.\text{-}3)$$

where the source function (ergiebigkeit) S_{iq} can be obtained from the corresponding expressions of the one-particle approximation. The absorption tensor K_{iqlm} is derived by an expansion of I_{iq} after eigenvectors. The tensor R_{iqlm} represents the FARADAY effect. An estimation of the single tensor components of these tensors has been made by KAWABATA (1964), SAZONOV and TSYTOVICH (1968) and others. However, as it was pointed out by ZHELEZNYAKOV (1968), the expressions for the components of the tensor K_{iqlm} derived by KAWABATA (1964) are only applicable if the radiation of each single electron corresponds to the polarization of one of the types of normal waves (e.g. absorption of cyclotron radiation in direction of the external magnetic field). In contrast to this the equations (3.4.3.-1)—(3.4.3.-3) are capable to describe any case of radiation in an anisotropic plasma.

3.5. Synchrotron radiation

3.5.1. Gyro-synchrotron radiation from electrons of medium energy

For many cases of solar radio bursts of moderate magnitude gyro-synchroton radiation of electrons of medium energy ($\beta = 0.25$—0.9) has to be taken into account. Basing on the results of SCHWINGER (1949), an evaluation of the emission and absorption coefficients and other radiation characteristica in this energy range has been carried out by TAKAKURA (1960) for circular and helical electron orbits. We start immediately with the study of arbitrary helical orbits $\Phi = 0 \ldots \pi/2$ which are for solar radio astronomy of great importance.

The absorption coefficient (not fully satisfying) calculated after the method of the EINSTEIN coefficients (TWISS, 1958) is given by

$$\varkappa_{\nu,s} = \sum_s - \frac{dN_s(\beta)}{dE} \cdot \frac{1}{2\,N_s(\beta)} \left(\frac{c}{\nu}\right)^2 \eta_{\nu,s}(\beta) \,, \qquad (3.5.1.\text{-}1)$$

where

$$\eta_{\nu,s}(\beta) = N_s(\beta) \frac{\omega_H^2 e^2}{2 \pi c} \frac{(\beta \sin \Phi)^2 (1 - \beta^2)}{(1 - \beta \cos \Phi \cos \theta)^4} \left\{[I'_s(s \alpha)]^2 + \left(\frac{\cos \theta - \beta \cos \Phi}{\beta \sin \Phi \sin \theta}\right)^2 [I_s(s \alpha)]^2\right\}. \tag{3.5.1.-2}$$

ω_H and α are given by the equations (3.4.1.-1) and (3.4.1.-12). It was taken $n = 1$.

The axis ratio of the polarization ellipse is represented by

$$R_{j,s} = - \frac{\cos \theta - \beta \cos \theta}{\beta \sin \Phi \sin \theta} \cdot \frac{I_s(s \alpha)}{I'_s(s \alpha)} . \tag{3.5.1.-3}$$

From (3.5.1.-3) it can be concluded that in general the occurrence of elliptical polarization is to be expected. Only for very special cases purely circular or even linear polarization will be obtained.

In order to calculate the ratio of the emission of the ordinary to the extraordinary component (or also total emission), TAKAKURA (1961) used the expression (3.3.2.-13) of the magneto-ionic theory leading under the assumption $X \ll 1$ to

$$R_{1,s} = \zeta \left[1 \mp \left(1 + \frac{1}{\zeta^2}\right)^{1/2}\right] \tag{3.5.1.-4}$$

with

$$\zeta = \frac{Y \sin^2 \theta}{2 \cos \theta} = \frac{(1 - \beta \cos \Phi \cos \theta) \sin^2 \theta}{2 s (1 - \beta^2)^{1/2} \cos \theta} . \tag{3.5.1.-5}$$

Hence for the fundamental frequency $s = 1$ which for solar conditions often is below the local plasma frequency, the following ratio of the emission of the ordinary component to the total emission is obtained

$$\gamma_{s=1} = \frac{(R_{j,1} - R_{1,1})^2}{(1 + R_{j,1})^2 (1 + R_{1,1})^2} , \tag{3.5.1.-6}$$

where

$$R_{1,1} = - \frac{\cos \theta - \beta \cos \Phi}{\beta \sin \Phi \sin \theta} \frac{I_1(\alpha)}{I'_1(\alpha)} . \tag{3.5.1.-7}$$

Tables of numerical values of $\gamma_{s=1}$ in dependence of β, θ, and Φ have been given by TAKAKURA (1961) and BÖHME (1964).

The energy losses by gyro-synchrotron radiation and collisions can be used for an estimation of the life-times of the emission as function of the strength of the magnetic field, particle energy, and particle density. A computation of the emission times Δt_{syn} and Δt_{coll} made by TAKAKURA (1961) based on the relations

$$\Delta t_{\text{syn}} = \frac{3 m_e c^2}{8 \pi^2 e^2} \frac{1}{\nu_H^2} \frac{(1 - \beta^2)^{1/2}}{\beta^2} [1 - (1 - \beta^2)^{1/2}] \quad [\text{s}] \tag{3.5.1.-8}$$

and

$$\Delta t_{\text{coll}} = 0.16 \frac{m_e^2 c^2}{\pi e^4} \frac{[1 - (1 - \beta^2)^{1/2}]^2}{N_e \beta (1 - \beta^2)} \quad [\text{s}] \tag{3.5.1.-9}$$

shows that the energy dependence of the emission time for collisions is stronger expressed than for synchrotron radiation (cf. fig. 11). For helical orbits of the emitting electrons the emission time is increased according to

$$[\Delta t_{\text{syn}}]_{\Phi \neq \frac{\pi}{2}} = \frac{1}{\sin^2 \Phi} [\Delta t_{\text{syn}}]_{\Phi = \frac{\pi}{2}} . \qquad (3.5.1.\text{-}10)$$

3.5.2. *Synchrotron radiation from highly relativistic electrons*

For electron energies $\beta > 0.7$ and $\nu \gg \nu_H$ from (3.4.1.-10) or (3.4.1.-22) the following approximation can be obtained

$$W_\nu(E, H, \Phi) = \frac{3^{3/2} c^2}{2 r_H} \left(\frac{E}{m_e c^2}\right)^4 \frac{\nu_H \nu}{\nu_c^2} \sin \Phi \int_{\nu/\nu_c}^{\infty} K_{5/3}(z)\, dz$$

$$= \frac{2 \sqrt{3}\, \pi e^2}{c} \nu_H \sin \Phi \frac{\nu}{\nu_c} \int_{\nu/\nu_c}^{\infty} K_{5/3}(z)\, dz$$

$$= \frac{4 \pi e\, m_e c^3}{\sqrt{3}\; E^2} \nu \sin \Phi \int_{\nu/\nu_c}^{\infty} K_{5/3}(z)\, dz$$

$$= \frac{\sqrt{3}\, e^3}{m_e c^2} H \frac{\nu}{\nu_c} \sin \Phi \int_{\nu/\nu_c}^{\infty} K_{5/3}(z)\, dz \qquad (3.5.2.\text{-}1)$$

(Schwinger, 1949; Oort and Walraven, 1956).

Here

$$\nu_c = \frac{3 c}{4 \pi r_H} \left(\frac{E}{m_e c^2}\right)^3 \sin \Phi = \frac{3}{2} \nu_H \sin \Phi \left(\frac{E}{m_e c^2}\right)^2 \qquad (3.5.1.\text{-}2)$$

denotes the critical frequency of the synchrotron emission which is closely related to the position of the spectral maximum of $W_\nu(E, H, \Phi)$. $K_{5/3}$ is the modified Hankel function. Comparing the results of different authors it must be noted, that different definitions of ν_c have been used. Sometimes the constant factor in (3.5.2.-2) is taken 1 or 3 (Ginsburg, 1951; Jackson, 1962; cf. Priester and Rosenberg, 1965).

The radiation is emitted into an effective solid angle $4 \pi \delta \sin \Phi$ around $\theta \approx \Phi$. For high energies and not too small values of θ the angle δ is small, but the approximation $\theta \approx \Phi$ becomes incorrect for small values of θ since then the contribution of low-order harmonics of the gyrofrequency becomes larger (cf. Takakura, 1967).

Numerical values of the function $F(y) = y \int_y^\infty K_{5/3}(z)\, dz$ have been calculated e.g. by Vladimirsky (1948) and Oort and Walraven (1956). For numerical studies also the following approximations are useful

$$\nu_c \gg \nu \left(\frac{\nu}{\nu_c} \lesssim 10^{-2}\right)$$

$$W_\nu = 2^{3/2}\,\Gamma\left(\frac{2}{3}\right)\frac{\sqrt{3}\,e^3}{m_e\,c^2}\,H\left(\frac{\nu}{\nu_c}\right)^{1/3}\sin\Phi = 5.16\cdot 10^{-22}\,H_{[\text{gauss}]}\left(\frac{\nu}{\nu_c}\right)^{1/3}\sin\Phi \quad [\text{cgs}] \tag{3.5.2.-3}$$

$$\nu_c \ll \nu \left(\frac{\nu}{\nu_c} \gtrsim 10\right)$$

$$W_\nu = \left(\frac{3\,\pi}{2}\right)^{1/2}\frac{e^3}{m_e\,c^2}\,H\left(\frac{\nu}{\nu_c}\right)^{1/2}\sin\Phi\,e^{-\nu/\nu_c}$$

$$= 2.93\cdot 10^{-22}\,H_{[\text{gauss}]}\left(\frac{\nu}{\nu_c}\right)^{1/2}\sin\Phi\,e^{-\nu/\nu_c} \quad [\text{cgs}] \tag{3.5.2.-4}$$

(JÄGER and WALLIS, 1961).

The shapes of the spectra $W_\nu(E, H, \Phi = \pi/2)$ are shown in fig. 20.

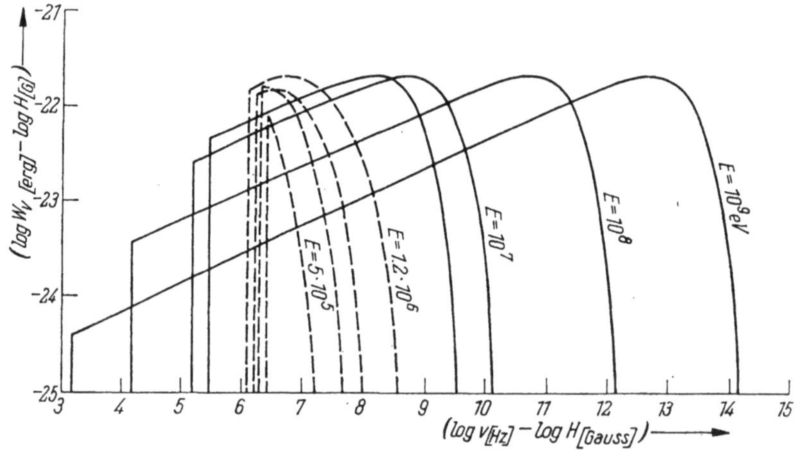

Fig. 20. Single particle synchrotron emission spectra after JÄGER and WALLIS, 1961

3.5.3. *The influence of the medium on the synchrotron radiation,* RAZIN *effect*

In addition to the elementary emission processes described by the one-particle approximation an influence of the medium is present in different ways expressed e.g. by the existence of a refractive index $n_j \neq 1$. One of the consequences of n_j consists of the action of refraction. We have also considered the relation between the (complex) refractive index and the absorption coefficient and the FARADAY effect. Now the influence of the medium on the synchrotron emission is of interest, rather than the consequences on the ray paths etc.

We consider a medium for which

$$0 < 1 - n_j \ll 1 \tag{3.5.3.-1}$$

is valid.

Then equation (3.5.2.-1) is written

$$W_\nu(E, H, \Phi, n_j) = \frac{2\sqrt{3}\,\pi\, e^2 \nu_H \sin\Phi}{c}\left[1 + (1 - n_j^2)\left(\frac{E}{m_e c^2}\right)^2\right]^{-1/2} \cdot \frac{\nu}{\nu_{c,n}} \int\limits_{\nu/\nu_{c,n}}^{\infty} K_{5/3}(z)\, dz \qquad (3.5.3.\text{-}2)$$

whereas

$$\nu_{c,n} = \nu_c \left[1 + (1 - n_j^2)\left(\frac{E}{m_e c^2}\right)^2\right]^{-3/2}. \qquad (3.5.3.\text{-}3)$$

Evidently modifications of the spectrum due to the presence of $n_j \neq 1$ become essential mainly at lower frequencies. Thus also the position of the spectral maximum of (3.5.2.-1) is changed, which is determined by the critical frequency. For $n = 1$ this position is approximately given by

$$\nu_{\max} \approx \frac{\nu_H}{2}\left(\frac{E}{m_e c^2}\right)^2 \sin\Phi\,. \qquad (3.5.3.\text{-}4)$$

On the other hand, for $n_j \neq 1$ we have

$$\nu_{\max,n} \approx \sqrt{2}\, \nu_p \left(\frac{E}{m_e c^2}\right). \qquad (3.5.3.\text{-}5)$$

From this a shifting of the spectral maximum towards higher frequencies and a suppression of the intensity especially at low frequencies in comparison to the case of $n = 1$ follow.

The suppression of the spectrum of the synchrotron radiation at longer wavelengths is known as RAZIN-effect or RAZIN-TSYTOVICH-effect (TSYTOVICH, 1951; RAZIN, 1960). As stated by several authors (ZHELEZNYAKOV, 1966; RAMANATY and LINGENFELTER, 1967; BOISCHOT and CLAVELIER, 1967; SIMON, 1969b) this effect may be of importance for the interpretation of the spectral features of certain type IV burst components.

According to (3.5.3.-2) the influence of the medium becomes essential if the condition

$$1 - n_j^2 \gtrsim \left(\frac{m_e c^2}{E}\right)^2 = 1 - \beta^2 \qquad (3.5.3.\text{-}6)$$

is fulfilled. If in accordance with (3.5.3.-1) for the refractive index in (3.5.3.-6) the simple expression $n^2 = 1 - X$ (table 6) is substituted, one obtains

$$\nu^2 \lesssim \nu_p^2 \left(\frac{E}{m_e c^2}\right)^2 = \frac{2\,\nu_p^2\, \nu_c}{3\,\nu_H \sin\Phi} \qquad (3.5.3.\text{-}7)$$

or with the aid of (3.5.3.-5)

$$\nu \lesssim \frac{\nu_{\max,n}}{\sqrt{2}}\,. \qquad (3.5.3.\text{-}8)$$

By putting ν approximately equal to ν_c GINSBURG and SYROVATSKY (1963, 1965) obtained the criterium

$$\nu \lesssim \nu_R = \frac{4\,e\,c\,N_e}{3\,H\sin\Phi} = \frac{\nu_p}{\alpha'} = \frac{3}{2}\frac{\nu_H}{\alpha'^2} \approx 20\,\frac{N_e}{H\sin\Phi} \quad \text{with} \quad \alpha' = \frac{2}{3}\frac{\nu_p}{\nu_H\sin\Phi} \tag{3.5.3.-9}$$

for a determination of the frequency range, in which the RAZIN effect cannot be neglected.

However, estimating N_e and H after (3.5.3.-9) at best only the order of magnitude could be obtained because the relation $\nu/\nu_c = 1$ as well as the use of the refractive index of the isotropic plasma appear as very rough approximations (cf. eg. BOISCHOT and CLAVELIER, 1967).

Synchrotron emission spectra of monoenergetic electrons for various values of the parameter $\alpha'\gamma$ have been calculated by RAMANATY and LINGENFELTER (1967) and RAMANATY (1968) which are shown in fig. 21. For very large values of $\alpha'\gamma$ the influence of the medium can be neglected, whereas in the solar corona values of α' typically smaller than 0.5 can be expected. α' diminishes with with increasing height above the photosphere.

For $n_j \approx 1$ and $\alpha' \approx 1$, i.e. $\nu_H \approx \nu_P \approx \nu_R$ a limiting case of the RAZIN effect is reached. Here the suppression of the synchrotron radiation occurs at frequencies where the medium has a significant influence on the propagation of the radio waves finally leading to the cutoffs at the plasma frequency.

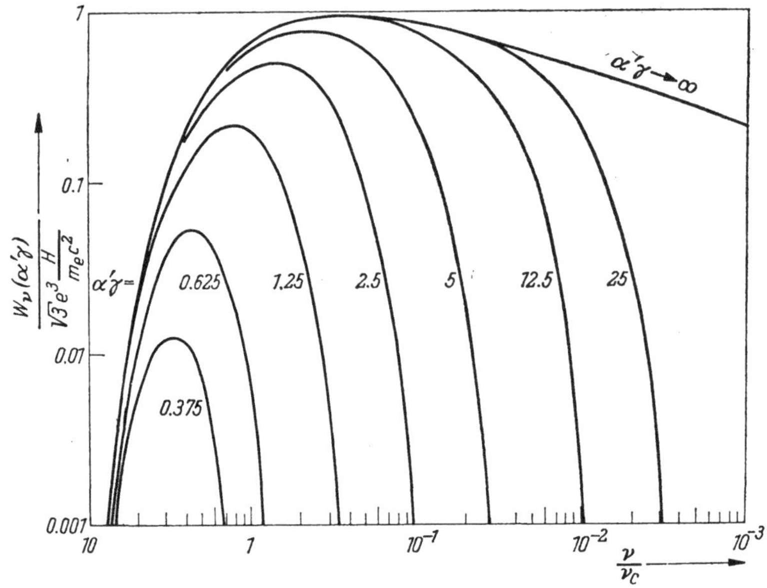

Fig. 21. Single particle synchrotron emission spectra for various values of the parameter $\alpha'\gamma$ after RAMANATY and LINGENFELTER (1967)

3.5.4. Polarization characteristics and synchrotron reabsorption

In order to obtain the intensity of the radiation the formulae of the single particle processes must be integrated with respect to the energy distribution of the emitting particles. In particular we get from (3.5.2.-1)

$$I_\nu = \frac{\sqrt{3}\, e^3}{2\pi R\, m_e\, c^2}\left[\iint N_e(E, R, \Phi)\, dE\, dV \cdot H \sin\Phi \frac{\nu}{\nu_c} \int_{\nu/\nu_c}^{\infty} K_{5/3}(z)\, dz\right] \tag{3.5.4.-1}$$

where $dV = R^2\, dR\, d\Omega$ is a volume element of the emission region (R — distance to the observer).

Assuming a distribution function in the following form

$$N_e(E, R, \Phi)\, dE = N_e(E, \bar{R}, \Phi) = K\, E^{-\gamma'}\, F(\bar{R}, \Phi)\, dE \tag{3.5.4.-2}$$

it follows from (3.5.4.-1)

$$I_\nu = \frac{\sqrt{3}}{4\pi}\frac{e^3}{m_e\, c^2}\frac{K}{R^2}\left(\frac{3\, e}{4\pi\, m_e^3\, c^5\, \nu}\right)^{\frac{\gamma'-1}{2}} \int_{y_2}^{y_1} F(\bar{R}, \Phi)\, dV\, (H \sin\Phi)^{\frac{\gamma'+1}{2}}$$

$$\cdot \int_{y_2}^{y_1} y^{\frac{\gamma'-1}{2}}\, dy \int_{y}^{\infty} K_{5/2}(z)\, dz\, . \tag{3.5.4.-3}$$

If

$$y_1 = \left[\frac{\nu}{\nu_c}\right]_1 = \frac{4\pi\, m_e^3\, c^5\, \nu}{3\, e\, H \sin\Phi\, E^2} \gg 1\, ; \qquad y_2 \ll 1\, . \tag{3.5.4.-4}$$

I_ν and the other Stokes parameters are obtained as

$$\left.\begin{aligned}
I_\nu &= A(\gamma')\frac{\gamma' + \frac{7}{3}}{\gamma' + 1}\cdot\frac{K}{R^2}\nu^{\frac{1-\gamma'}{2}}\int (H \sin\Phi)^{\frac{\gamma'+1}{2}} F(\bar{R}, \Phi)\, dV\, ,\\
Q_\nu &= A(\gamma')\frac{K}{R^2}\nu^{\frac{1-\gamma'}{2}}\int (H \sin\Phi)^{\frac{\gamma+1}{2}} F(\bar{R}, \Phi)\cos 2\chi\, dV\, ,\\
U_\nu &= A(\gamma')\frac{K}{R^2}\nu^{\frac{1-\gamma'}{2}}\int (H \sin\Phi)^{\frac{\gamma'+1}{2}} F(\bar{R}, \Phi)\sin 2\chi\, dV\, ,\\
V_\nu &= 0\, ,
\end{aligned}\right\} \tag{3.5.4.-5}$$

where

$$A(\gamma') = \frac{\sqrt{3}\cdot 2^{\frac{\gamma'-3}{2}}}{4\pi}\,\Gamma\left(\frac{3\gamma' - 1}{12}\right)\Gamma\left(\frac{3\gamma' + 7}{12}\right)\cdot\frac{e^3}{m_e\, c^2}\cdot\left(\frac{3\, e}{4\pi\, m_e^3\, c^5}\right)^{\frac{\gamma'-1}{2}} \tag{3.5.4.-6}$$

(cf. Korchak, 1963).

$\varkappa_\nu$ can be calculated from I_ν applying the method of EINSTEIN coefficients:

$$\varkappa_\nu = \frac{1}{I_\nu}\frac{dI_\nu}{ds} = -\frac{c^2}{8\,\pi\,\nu^2}\int\limits_0^\infty \frac{\partial N_e(E, R, \Phi)}{\partial E} W_\nu(E, H, \Phi)\,dE \qquad (3.5.4.\text{-}7)$$

and for an energy spectrum of the form (3.5.4.-2) one obtains

$$\varkappa_\nu = B(\gamma')\frac{e}{6\,\pi\,m_e}\left(\frac{3\,e}{2\,\pi\,m_e^3\,c^5}\right)^{\gamma'/2} K(H\sin\Phi)^{\frac{\gamma'+2}{2}}\,\nu^{-\frac{\gamma'+4}{2}} \qquad (3.5.4.\text{-}8)$$

with

$$B(\gamma') = \frac{3\sqrt{3}}{4}\frac{\gamma'}{\gamma'+2}\,\Gamma\left(\frac{3\,\gamma'+2}{12}\right)\Gamma\left(\frac{3\,\gamma'+22}{12}\right) \qquad (3.5.4.\text{-}9)$$

(GINSBURG and SYROVATSKY, 1963).

In the case of small optical depths the dependence on the frequency for synchrotron emission with weak reabsorption it follows

$$I_{\nu,\,\text{syn}} \sim \nu^{\frac{1-\gamma'}{2}}\,\nu^{\frac{4+\gamma'}{2}} = \nu^{5/2}\,. \qquad (3.5.4.\text{-}10)$$

Thus in this case no dependency of the exponent γ' on the energy distribution can be stated.

3.5.5. *Negative reabsorption and coherent synchrotron emission*

Emission mechanisms which are similar to the action of masers and lasers are based on the effect of a negative absorption coefficient and therefore emit "coherent radiation" in that sense used e.g. by ZHELEZNYAKOV (1964). They have been proposed for an interpretation of the occurrence of high intensities of solar radio bursts (type IV bursts).

First attempts in a discussion of the negative absorption with respect for the radio bursts have been made by TWISS (1958). The role of the energy distribution function for the occurrence of the effect of negative reabsorption has been considered in more detail by SMERD (1965) and ZHELEZNYAKOV (1967).

From the formula of the synchrotron absorption coefficient (3.5.4.-7) one obtains by partial integration

$$\varkappa_\nu = \frac{c^2}{8\,\pi\,\nu^2}\int\limits_0^\infty \frac{N_e(F)}{E^2}\frac{d}{dE}[E^2\,W_\nu(E)]\,dE\,. \qquad (3.5.5.\text{-}1)$$

According to (3.5.3.-6) two extremal regions

$$\frac{\left(\frac{\nu_p}{\nu}\right)^2}{\left(\frac{E}{m_e\,c^2}\right)^2} \gg 1 \qquad (3.5.5.\text{-}2)$$

and

$$\frac{\left(\frac{\nu_p}{\nu}\right)^2}{\left(\frac{E}{m_e c^2}\right)^2} \ll 1 \qquad (3.5.5.\text{-}3)$$

are possible, for which the influence of the reabsorption becomes essential or neglectable, respectively. In the case of (3.5.5.-3) (vacuum!) for arbitrary energy distributions always a positive $\varkappa_\nu$ is resulting, so that here only the case (3.5.5.-2) is of interest.

From (3.5.3.-6) and (3.5.5.-1) it follows immediately, that the following necessary conditions must be simultaneously fulfilled

$$\frac{\partial N_e(E)}{\partial E} > 0 \; ; \quad \frac{\partial W_\nu(E)}{\partial E} < 0 \qquad (3.5.5.\text{-}4)$$

if a negative absorption effect could be possible. The first condition denotes an overpopulation of higher energy levels and therefore thermal non-equilibrium. This may be realized e.g. in electron beams penetrating the solar atmosphere. The second condition depends on the special emission mechanism and is realized beyond others in the occurrence of "resonance peaks" with respect to E.

In contrast to a result of SMERD (1965), who neglected the influence of the medium, in a plasma the synchrotron radiation can be amplified by negative reabsorption. Such radiation instabilities ("coherent" emission) have been treated by FUNG (1968, 1969) and ZHELEZNYAKOV (1966). Up to now, basing on instability criteria only qualitative conclusions on the burst spectrum are possible. For a quantitative computation of the emission amplitudes non-linear calculations of negative absorption processes are necessary which at present are not yet well developed.

Finally it should be generally remarked, that the considerations given here are not necessarily restricted to electrons as emitting particles. In principle all formulae could be extended also to positrons or other charged particles, if needed.

3.6. Čerenkov radiation

3.6.1. Coherent electromagnetic waves

We consider the x-component of the radiation field vector $\boldsymbol{E}$ of an electron moving with a speed $\boldsymbol{v}_e$ in x-direction:

$$E_x = \int E_{k_{j,x}} \exp\{i\, k_{j,k}\,(x - v_{k_{j,x}}\, t)\}\, dk_{j,x} \, . \qquad (3.6.1.\text{-}1)$$

Here the coherence condition for the occurrence of the ČERENKOV-effect follows

$$\boldsymbol{v}_e = \boldsymbol{v}_{k_j} \qquad (3.6.1.\text{-}2)$$

or

$$\omega = \boldsymbol{k}_j \cdot \boldsymbol{v}_e = k_j v_e \cos \psi \quad \text{with} \quad \psi = \sphericalangle (\boldsymbol{k}_j, \boldsymbol{v}_e) \tag{3.6.1.-3}$$

or using

$$n_j = \frac{c\, k_j}{\omega}, \qquad \beta = \frac{v_e}{c} \tag{3.6.1.-4}$$

one obtains

$$\cos \psi = \frac{1}{\beta_e n_j}. \tag{3.6.1.-5}$$

This means that the ČERENKOV effect becomes possible, if

$$\beta_e n_j \geqq 1 \tag{3.6.1.-6}$$

i.e. if the speed of the electron v_e reaches the phase velocity v_{k_j} of the electromagnetic waves in the medium. In an anisotropic medium the ČERENKOV radiation can be regarded theoretically as limiting case of the synchrotron radiation for

$$H \to 0 \; (r_H \to \infty) \quad \text{and} \quad \beta_e \cos \Phi \cdot n_j \cos \theta > 1 \tag{3.6.1.-7}$$

(anomalous DOPPLER effect (3.4.1.-9)).

For a not vanishing external magnetic field especially the components of the speed along the field ($\Psi = \theta$) are of interest for a study of the ČERENKOV radiation, while in the case of synchrotron emission the components perpendicular to the field direction are most important.

From the dispersion relation (3.3.1.-1) it becomes evidient, that the condition (3.6.1.-4) for electromagnetic waves (contrary to plasma waves!) and arbitrary $\beta < 1$ cannot be realized in an isotropic plasma because $n_j \leqq 1$. For an anisotropic plasma, however, it is seen from (3.3.1.-5), that under special circumstances ČERENKOV radiation can be emitted in the extraordinary mode. For $Z \neq 0$ the ČERENKOV losses are determined by the real part of n_j^2.

The regions characterized by ν_H, ν_p, and ν_{coll} in which ČERENKOV radiation takes place have been discussed e.g. by SOLLFREY and YURA (1965). In this paper the reliability of the often cited calculations of KOLOMENSKY (1956) have been called in question. However, here we can cut it short, because in most cases of interest for solar radio astronomy the part of the ČERENKOV radiation in the extraordinary electromagnetic wave mode is far below the amount of synchrotron radiation. In this respect also the paper of MARSHALL (1956) in which the attempt was made to explain most of the features of nonthermal radio emission by the ČERENKOV radiation was too optimistic and not correct. Nevertheless this paper gave an initial impuls for a discussion of the usefulness of the ČERENKOV effect with regard to solar problems.

3.6.2. Plasma waves and L—T interactions

For the reasons given in the preceding section the ČERENKOV effect is far more effective for plasma waves than for the electromagnetic waves. For longitudinal waves the condition $n^2 < 1$ is realized e.g. for $\omega < \omega_p$ and a coherent ČERENKOV plasma radiation is expected in a wide range where $v_{\text{th}} < v_{k_j} \leqq v_e$.

The lower limit of the last inequality is based on the fact, that a propagation is only possible for the waves which satisfy the condition

$$\frac{\lambda}{2\pi} \geqq D\,, \quad \text{where} \quad D = \left(\frac{kT}{4\pi e^2 N_e}\right)^{1/2}. \tag{3.6.2.-1}$$

The emitted energy has been estimated as

$$dW \leqq \int\limits_{\omega_1}^{\omega_2} \frac{e^2\,\omega}{v_e\,n^2}\,d\omega = \frac{e^2\,\omega_p^2}{2\,v_e} \ln\left(\frac{2\,v_e^2}{3\,v_{\text{th}}^2}\right) \tag{3.6.2.-2}$$

(ZHELEZNYAKOV, 1964; COHEN, 1961).

Hence it follows that the energy content of the plasma waves under the condition of the solar atmosphere exceeds that of the electromagnetic waves by several orders of magnitude. But plasma waves can propagate only in a medium so that they cannot leave the solar atmosphere and are detectable on the earth only if a conversion into electromagnetic waves by some transformation processes takes place.

There are various types of plasma waves as electronic and magnetic plasma waves, the ALFVÉN waves, the magneto-acoustic waves etc. The main types and modes are contained in the CMA-diagram of fig. 13. In a homogeneous plasma and within the range of a linear approximation these wave types as well as the transversal wave modes are propagating independently of each other without any interaction. However, in an inhomogeneous medium or in the framework of a non-linear theory a coupling and transformation of energy between different wave modes (and also between waves of one and the same mode but different wave numbers) occur. Thus a study of solar plasma wave phenomena involves the problems of the generation of the different types of these waves and the consideration of the mechanisms of their convertion into the transversal wave modes.

Concerning the generation of the plasma waves we note briefly that the ČERENKOV effect gives only one possibility for the production of waves and also other emission mechanisms are thinkable. As a matter of fact the hot and warm plasma is rich in producing various kinds of turbulence which are identical with the occurrence of plasma waves.

If the existence of sufficient plasma waves in the solar corona is presumed, the problems of the transformation of longitudinal into transversal wave modes is of great interest, which here can be touched only briefly. Mathematically the occurrence of wave-wave interactions is characterized by the fact that under special circumstances the dispersion relation cannot be longer factorized as it was presumed in most cases of the foregoing sections and a mode coupling arises.

Roughly speaking, there are three large categories of wave-wave interactions of interest which are schematically compiled in table 10. The first group refers to macroscopic inhomogenieties of the medium (e.g. stratified density profiles,

Tab. 10. Classification scheme of wave-wave interactions

(L-T-Transformation)

Macroscopic inhomogeneous medium	Scattering at microscopic inhomogenieties	Coupling at critical regions of the dispersion relation
– Gradient coupling – Magnetic field coupling – Drift coupling etc.	– RAYLEIGH scattering – Combination sattering (thermal or non-thermal irregularities, electrons, and ions)	– Resonance phenomena, efficient by collisions – Transitions to other parts of a dispersion curve by changing plasma parameters

($\triangleq$ non-linear effects in a warm plasma, in general)

gradients of magnetic fields etc.) influencing the propagation characteristics so that wave-wave interactions become possible. The second type of coupling processes is due to the action of microscopic irregularities and refers to scattering processes as RAYLEIGH scattering and combination scattering. The third type refers to such regions of the dispersion relation where different branches of the the dispersion curves cameclose together so that the equality of the wave number k_i of single modes provides a necessary (but not sufficient!) condition for the possibility of the occurrence of a mode coupling. First attempts for an interpretation of solar radio emissions by this effect were made by DENISSE (1960). But also simply transitions to different parts of one and the same branch of a dispersion curve appear possible.

Because there is a dependence of the degree of some L–T-transformations on the expression $f(\lambda/L)$ (λ—wavelength, L—characteristic dimension of the inhomogenieties etc.) only a relatively weak efficiency of this transformation in the order of $10^{-5}...10^{-7}$ must be expected. If interactions in form of a spontaneous conversion of the plasma waves into electromagnetic wave modes are proposed, an appreciable energy content of the plasma waves would be required. Above a certain limit of the observed radio emission ($\triangleq T_e \sim 10^9$ °K) the mechanism of induced emission must be invoked. But these things up to now are not well investigated. A summary of the role of plasma radiations in astrophysics has been given by KAPLAN and TSYTOVICH (1968).

As discussed in the last chapter, possibly the continuum of the type IVmB bursts can be interpreted by the action of a transformation of longitudinal into transversal waves. Other important applications of mode coupling in solar radio astronomy are given by the phenomena of drift bursts, especially the type III bursts (GINSBURG and ZHELEZNYAKOV, 1959; ZHELEZNYAKOV and ZLOTNIK, 1962; ZAJTSEV and KAPLAN, 1966) ZHELEZNYAKOV and ZAJTSEV, 1970).

3.7. Acceleration mechanisms and particle energy spectrum

3.7.1. Basic problems of the origin of flares and bursts

The action of the emission mechanisms considered in the previous sections is based on the existence of accelerated particles. Evidently strong and rapid acting acceleration processes are required to explain the suddeness and violence of the observed burst phenomena. In looking for the origin of the accelerated particles, the basic, and since more than hundred years unsolved, question of the origin of the flare-burst phenomenon is touched. Considering this problem, the interpretation of the flare-burst radiation in the whole electromagnetic spectrum ranging from the gamma- to the lowest radio waves is linked with the exploration of the solar particle emission and leads to similar problems as arising from the question of the origin of the non-solar cosmic radiation.

In principle the following mechanisms are known resulting in a production of accelerated particles:

1. Acceleration by electromagnetic fields
   a) Fermi-Parker mechanism
   b) Betatron process and related mechanisms
   c) Direct dissipation of magnetic energy
2. Acceleration by mechanical forces
   a) Energy dissipation in shock waves
   b) (Adiabatic) compression
   c) Radiation pressure
   d) Gravitational forces
3. Production of energetic particles by nuclear reactions

The last point describes no particle acceleration in the usual sense, but for completion nuclear reactions are involved in a discussion of potential sources of fast particles, though this may account only for rare cases of big events.

It is evident that the question of finding the actual acceleration process is closely connected with the problem of the origin of the energy of the flare-burst phenomenon in which especially two difficulties must be envisaged.

One of these basic difficulties arise from the magnitude of the energy exchange covering large volumina in the solar atmosphere. The second main difficulty consists in the explanation of the limited duration of this energy exchange process. The energy released in large flares has been estimated unanimously up to the order of 10^{33} erg. In comparison to this figure the kinetic energy content of the whole solar atmosphere amounts to only $5 \cdot 10^{31}$ erg. Nevertheless the magnetic energy of a large solar spot group (assumed as potential field) is of the order of 10^{34} erg and therefore principially can be regarded as a potential source of the flare-burst energy. For this reason most attempts of an interpretation of the origin of the flare-burst phenomenon are refering to magnetodynamical processes as a direct or indirect cause of the particle acceleration.

Besides the question after the origin of the flare energy the questions after the location and nature of the energy storage and the initial process of the energy release (trigger mechanism, kind of instability) demand a solution. At first one has to decide, whether a subphotospheric origin of the flare process (which would be intimately merged with problems of the origin of solar activity in general) is accepted or not. For the reasons that photospheric disturbances are not observed during flares and that a subphotospheric origin would only shift the problem into deeper levels of the Sun without hope for a prompt solution, most authors ruled out the discussion of subphotospheric sources until now. Therefore, mainly the questions of potential flare processes situated above the photosphere will be considered here.

For a long time problems arised with respect to the dissipation time of the magnetic fields, which is according to the well known relation of magnetohydrodynamics

$$t_d = 4\pi\sigma L^2 \tag{3.7.1.-1}$$

(σ — electrical conductivity, L — characteristic length of the emitting volumes).

Under plausible values for the flare volumina ($L \sim 10^{10}$ cm) equation (3.7.1.-1) leads to much too high dissipation times ($t_d \sim 10^3$ years). In order to obtain the required order of magnitude of maximally 10^3 sec for a given σ the dimensions of the flare volumen at least in one direction must be very small ($\sim$1 cm!). This led to the idea of thin discharge sheets near neutral points of the magnetic field which have been discussed at first by DUNGEY (1953) and later was considered in the SWEET-PARKER model (SWEET, 1958; PARKER, 1963). But still in this model the resulting dissipation times remained too high by about two orders of magnitude. Therefore other possibilities for a reducing of the quantities L or σ must be taken into consideration.

Different ways of a reduction of L have been proposed, e.g. by assumption of finite-conductivity instabilities (JAGGI, 1965), by the generation of filament structures (WENTZEL, 1963; COPPI, 1964), by invoking the action of magnetohydrodynamic shock waves (PETCHEK, 1964) or space-charge waves (SCARF et al., 1965).

Also a reduction of σ by assumption of a greater part of neutral particles (GOLD and HOYLE, 1960) does not essential improve the situation. Other attempts have been made using the tensor character of σ or modifying the basic equation (3.7.1.-1) (PIDDINGTON, 1967). At present the most promising proposals to obtain smaller values of t_d invoke the action of turbulence reducing σ and hence the dissipation times up to a factor of four orders of magnitude (STEENBECK et al., 1966).

However, in all these cases further difficulties arise from the fact that the position and spatial extent of the volumina with a maximal storage of energy do not necessarily bear an obvious relationship to the places of maximum energy exchange. Therefore a fast transfer of energy is required. SCHATZMANN (1967) invoked the action of heat transfer in order to account for the observed extension of the flare regions. He estimated the characteristical time of heat transfer t_h which is proportional to

$$t_h \sim \frac{L^2 N_e}{\sigma'} \qquad (3.7.1.\text{-}2)$$

as consistent with a heat flow of the order of $10^4 \cdots 10^6$ erg cm^{-2} which is of the order of energy liberated in a solar flare. Other proposals of an energy transfer by magnetohydrodynamic waves or fast particles have been discussed by NEUMAN (1967), WARWICK (1962), and ELLIOT (1969).

Concerning the initial process of the energy release a state of metastabile equilibrium is needed which leads to an instability and triggers the stored energy. The knowledge of this trigger mechanism appear essential for the solution of the central question of an understanding of the flare phenomenon, the explanation of the explosive phase in the flare development, during which up to about 10^{32} erg of stored energy are exchanged mainly into kinetic energy of mass motion. Though this problem is not yet solved a compilation of some instabilities potentially important for solar physics is given in tab. 11. A more detailed survey on this subject has been given by LEHNERT (1967).

Tab. 11. Instabilities

	Electrostatic	Electromagnetic
Macro-instabilities	Flute instability KELVIN-HELMHOLTZ instability Thermal instability	Sausage instability Collision-free shock instability Resistive tearing mode instability
Micro-instabilities	Two-stream instability Beam-plasma instability Ion-wave instability Cyclotron instabilities	Collisionless tearing mode instability ALFVÉN-wave instability Mirror instability

Concerning the action of the acceleration process itself it must be remembered that all hitherto proposed electromagnetic acceleration mechanisms including the FERMI mechanism (cf. 3.7.3) and the direct transformation of magnetic into kinetic energy (cf. 3.7.5) are more or less directly related to FARADAY's induction law. The production of an extended continuous spectrum of the particle energies is characteristical for this kind of acceleration mechanisms which is basically different from e.g. the limited interval of energies produced by nuclear reactions.

Other attempts of an explanation of the acceleration by the action of the radiation pressure (MUSTEL, 1966; WARWICK, 1957) or by a storage of kinetic energy (CARMICHAEL, 1964; STURROCK, 1968; ELLIOT, 1969) are also problematically. The main difficulty again consists in an satisfying explanation of the rapid increase of the equivalent temperature of at least two orders of magnitude within a few seconds of time.

Similarily the attempts basing on gravitational forces or other mechanisms are questionable. Interesting ideas in this respect are contained in the inversion model of STURROCK and COPPI (1964) which bases on the action of gravitational forces. This concept could hold at best as trigger mechanism. The energy content of the gravitational field is comparable with that of the kinetic energy in the solar atmosphere and therefore insufficient to account for the flare process. Also the observations of burst sources drifting outward into the interplanetary space makes such a model difficult.

After this general discussion of some of the main ideas on the nature of the particle acceleration in solar flares and bursts we try to discuss a few of the main acceleration mechanisms in some more detail.

3.7.2. Direct electric field acceleration, discharge theory

The first greater attempt of a theoretical model of a flare mechanism has been made on the basis of a discharge theory (for the first beginnings cf. e.g. SWANN, 1933). In this theory the acceleration of the charged particles is explained by means of electrical fields (GIOVANELLI, 1946—48). Several authors, however, pointed out the impossibility of the existence of the necessary high electric field strengths in an ionized medium. Therefore the action of electrostatic forces must be largely excluded (SWEET, 1958; PARKER, 1963) and the discharge theory in its original form could not be maintained. Nevertheless, the observational evidence for the existence of strong electric currents in solar flares is stimulating a theoretical discussion on this subject (WENTZEL, 1963; CARLQUIST, 1969).

Simple drifts of particles with the velocity v induced by changes of the magnetic field produce an electrical field $\boldsymbol{E} = -(\boldsymbol{v} \times \boldsymbol{H})$ which following LENZ' law does not contribute to the particle acceleration, when the betatron effect by LARMOR motion is neglected. This situation is changed if electrical fields due to dissipative processes occur.

Then it can be written

$$\boldsymbol{E}_d = \frac{\boldsymbol{j}}{\sigma} \qquad (3.7.2.\text{-}1)$$

which causes a drift velocity

$$\boldsymbol{v}_d = \frac{\boldsymbol{E}_d \times \boldsymbol{H}}{H^2}, \qquad |v_d| \sim \frac{1}{4\pi\sigma L} = 5 \cdot 10^{12}\, T^{-3/2}\, L^{-1} \quad [\text{cm s}^{-1}]\,. \qquad (3.7.2.\text{-}2)$$

This velocity is of the same order of magnitude as the drift velocity of the magnetic field lines. High velocities can be expected near the neutral points of the magnetic field where the particle acceleration takes place in approximately the same manner as in the magnetic field-free case. The E-components parallel to the magnetic field direction can be treated analogeously. DUNGEY (1958), DE JAGER (1960), and SWEET (1958) have shown, that electric fields of the order of 1 eV cm^{-1} can be expected. But this value is not sufficient to account for the main part of the accelerated particles.

3.7.3. The FERMI-PARKER *mechanism*

One of the most famous, in different variations often quoted acceleration mechanisms is based on the presumption of temporal changing magnetic fields "reflecting" charged particles running against them which was developed in its outlines by FERMI (1949). The reflection is due to the adiabatic invariancy of the magnetic moment

$$\mu' = \frac{m\, v^2}{2\, H} \tag{3.7.3.-1}$$

which excepting the factor $2\, m\, c/e$ is identical with the impuls p of the LARMOR motion (HERTWECK and SCHLÜTER, 1957):

$$\frac{p_T^2}{H} \equiv \frac{p^2 \sin^2 \Phi}{H} = \text{const}\,, \qquad p_L^2 \equiv p^2 \cos^2 \Phi = \text{const}\,. \tag{3.7.3.-2}$$

For a (for simplicity as static assumed) magnetic field the reflection condition given by FERMI (1954) is valid:

$$H > \frac{1}{q}\,, \qquad q = \frac{\sin^2 \Phi}{H}\,, \qquad (\Phi = \sphericalangle\, (\boldsymbol{v}, \boldsymbol{H}))\,. \tag{3.7.3.-3}$$

Correspondingly reflections at "magnetic walls" moving with a velocity w cause accelerations or decelerations of the reflected particles, if the components of $\boldsymbol{v}$ are directed oppositely or along the direction of w (headon or overtaking collisions), respectively.

According to FERMI (1949) the average gain in energy of a non-relativistic particle of the energy W after a collision with a magnetic wall is in first order

$$dW = \beta^2\, W\,. \tag{3.7.3.-4}$$

After n such collisions the energy becomes

$$W = m\, c^2 \exp\, (\beta^2\, n)\,. \tag{3.7.3.-5}$$

Taking into account the mean free path length Λ and the average time between two collisions τ, the energy of one particle at the time t is written

$$W(t) = m\, c^2 \exp \left\{\frac{\beta^2\, t}{\tau}\right\}. \tag{3.7.3.-6}$$

Then the probability for a particle of the "age" t to possess an energy between W and $(W + dW)$ is given by

$$\Gamma(W)\, dW = \frac{\tau}{\Lambda\, \beta^2}\, (m\, c^2)^{\frac{\tau}{\Lambda \beta^2}}\, W^{-(1+\tau)/(\Lambda \beta^2)}\, dW \tag{3.7.3.-7}$$

which yields an energy spectrum with a negative exponent.

Two basic conditions must be realized in order to make the Fermi process efficient. The first of them consists in the requirement of a permanent redistribution of the pitch angles Φ. Secondly, it must be required that the acceleration processes are prevailing over the deceleration processes which depend on the motion of the reflecting walls and the injection spectrum of the particles.

A repeated redistribution of the angles Φ without simultaneous deceleration is necessary for the maintenance of the acceleration mechanism because by the action of the FERMI process Φ decreases and therefore after a certain period the particles escape in direction of the magnetic field lines and do not suffer further reflections. PARKER (1958) proposed a solution of this problem by the assumption of narrow shock fronts of magnetohydrodynamic waves causing a reorientation of the particle velocities. This mechanism is only efficient, if the thickness of the shock front is less than the LARMOR radius of the particles. This condition cannot be expected to be fulfilled in every case. Therefore, TAKAKURA (1961) has proposed another mechanism of COULOMB collisions with thermal electrons which also produce a redistribution of the electron velocities.

As it is well known the second condition of the predominance of the acceleration over the deceleration processes was solved by PARKER (1958) by the assumption of crossing magnetohydrodynamic shock waves as reflecting agens. Here the assumption of shock waves is not essential for the acceleration process itself. A prevailing of the acceleration processes is also achieved by an existence of common ALFVÉN waves where the gain of velocity is

$$\Delta v_L = 2\, v_m\,, \qquad v_m = \frac{B_0}{\sqrt{4\,\pi\,\varrho_0}} \tag{3.7.3.-8}$$

(TAKAKURA, 1961; WENTZEL, 1963).

It should be noted that the FERMI-PARKER mechanism leads to the same velocities of protons and electrons, i.e. it works indenpedently of the mass of the accelerated particles. In this way, protons or ions obtain much more energy than electrons with the same initial velocity.

3.7.4. *Other electromagnetic acceleration mechanisms*

The FERMI-PARKER mechanism has the great disadvantage that it acts too slowly. Thus it could hardly explain the rapid primary acceleration in the explosive phase of a flare. The repeated collisions with the magnetic walls make it more appropriate to act as secondary acceleration mechanism, e.g. in second part events of the type IV burst complex. Moreover, it appears questionable that in all cases crossing shock fronts are present. Therefore the necessity arises to look for other more efficient mechanisms capable of explaining a faster acceleration.

Many authors have treated this problem without finding a completely satisfying solution. Usually the starting point is the idea that an increase of the magnetic field strength causes an enhancement of the kinetic energy of the LARMOR motion according to $p_T^2 \sim H$ as a consequence of the adiabatic invariancy of the magnetic moment of the LARMOR motion. This is applied e.g. in the betatron process, where the magnetic field strength in first approximation is taken proportional to the particle density and thus enables an estimation of the upper limit of the effiency of this mechanism (WENTZEL, 1964).

Estimations according to FARADAYS induction law

$$\left|\oint E_{\text{ind}}\, ds\right| = \frac{1}{c}\left|\frac{\delta}{\delta t}\int H\, d\Sigma\right|, \tag{3.7.4.-1}$$

(where the path s encloses the region Σ) have been made by DE JAGER (1965) and others. For the conditions found in solar flares an acceleration of electrons up to $1.5 \cdot 10^5$ km s^{-1} can be explained by this concept, if an average initial speed of 10 km s^{-1} is assumed. About the deeper origin of the flare phenomenon, however, no statements can be made in the framework of these considerations.

Also other modifications of electromagnetic acceleration processes, e.g. magnetic pumping gave no answer on the question of the flare origin. There is still a need for an explanation of an unknown primary acceleration to deliver the injection particles (cf. also e.g. SANDSTRÖM, 1965). Summarizing these results one has to conclude that the above mentioned electromagnetic acceleration processes as well as the FERMI-PARKER mechanism appear at best to account for the secondary acceleration process but not for the impulsive primary process taking place in the impulsive or flashphase of the solar flare-burst phenomenon.

3.7.5. *Direct transformation of magnetic into kinetic energy*

In the foregoing sections it was concluded that a fast acting and effective primary acceleration process is needed which probably takes its energy form the magnetic field. At present perhaps the most promising way to tackle this problemis in a combination of the acceleration process with dissipative processes of the magnetic field, i.e. with the direct transformation of magnetic energy into kinetic energy of the accelerated particles.

The hitherto known models of such a mechanism start from a study of the magnetohydrodynamic flow along a neutral line in a X- (or Y-) like configuration of the magnetic field (SYROVATSKY, 1966). The motion of the plasma near this neutral zone is calculated as a consequence of the assumption of small changes of the magnetic vector potential using the linearized magnetohydrodynamic equations. By this effect a field deformation occurs leading to a peculiar cumultative effect which results in the generation of a current sheet varying with time. At both sides of the field configuration streams of accelerated particles escape while in the central part matter flows perpendicular to the neutral line transfering and dissipating magnetic energy (fig. 22).

In this theory difficulties arise with respect to the validity of the assumption of slow changes of the magnetic moment of the spot field configuration. Further there arises the question whether it is justified that the energy dissipation is restricted only to regions in the vicinity of neutral point or lines. Finally, the

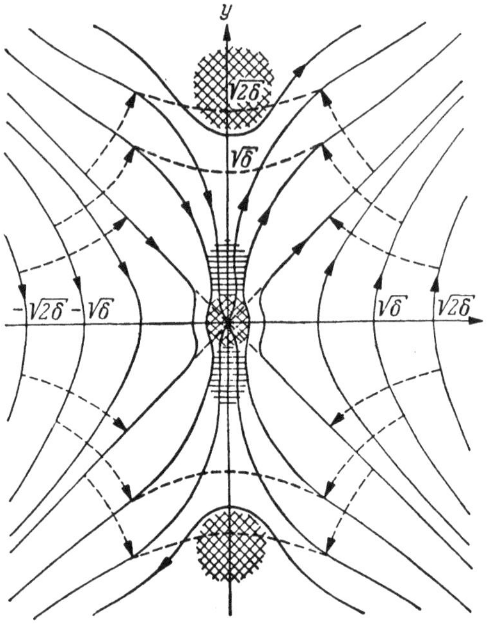

Fig. 22. MHD motion near a neutral line (y) by a displacement δ of the magnetic field configuration after SYROVATSKY (1966).
Hatched areas refer to compressions

basic question after the cause of the assumed changes of the vector potential remains open. Recently KŘIVSKY (1968) has extended the SYROVATSKY model with respect to a better describing of observational details of the optical flare development in order to explain the two-ribbon structure of proton flares and the generation of loops. Nevertheless, in spite of this progress up to now a theory satisfying in all respects and representing a non-linear, instationary, three-dimensional magnetohydrodynamic and plasma kinetic process is still missing.

3.7.6. Production of energetic particles by nuclear reactions

The idea that nuclear reactions may play somehow role in solar flares is far from being new. The preposition for an occurrence of nuclear processes is a sufficient strong injection spectrum of energetic particles. In this way, nuclear reactions at best play the role of secondary processes only and do not contribute directly to the solution of the origin of the flare phenomenon itself. Nevertheless the question of a participation of nuclear processes is important and, stimulated by observations of energetic solar protons and investigations about the spectrum of solar microwave bursts some discussions on this topic are to be quoted.

As it was stated by STEIN and NEY (1963), for an explanation of white-light flares by synchrotron radiation in magnetic fields of some 10^2 gauss, ultra-relativistic electrons with LORENTZ factors $\gamma = E/(m_e c^2)$ of 20 and more are required. In spite of this, the observed proton spectrum is only of the order of $\gamma = 1...2$. Such a discrepancy between the proton and electron energies would not be in accordance with the acceleration mechanisms discussed in the previous sections. A possible solution was proposed by RAMANATY and LINGENFELTER (1966) invoking the production of positrons from the decay of pions, which arise from the inelastic collisions of accelerated protons and other particles with the matter of the ambient plasma of the solar atmosphere. This can be described by the reaction

$$\pi^+ \rightarrow \mu^+ + \nu\,, \qquad \mu^+ \rightarrow e^+ + \nu + \bar{\nu} \tag{3.7.6.-1}$$

with reaction times of $2.6 \cdot 10^{-8}$ s and $2.2 \cdot 10^{-6}$ s.

The positive pions are generated by the following processes

$$H^1(p, n\,\pi^+)\,H^1 \tag{3.7.6.-2}$$

(treshold energy 287 MeV) and

$$H\,e^4(p, p\,n\,\pi^+ \cdots) \tag{3.7.6.-3}$$

(treshold energy 198 MeV).

According to ALFVÉN (1965) the life time of the generated positrons can be estimated as follows

$$\tau_d > \frac{10^{13}}{N_e}\ [\mathrm{s}]\,. \tag{3.7.6.-4}$$

Hence for e.g. $N_e = 10^9\ \mathrm{cm}^{-3}$ the life time is still 1–2 orders of magnitude greater than the emission time for synchrotron radiation which for $H = 300$ gauss and $\gamma_e = 20$ is at about 10^2 s.

The distribution function $F(\gamma_e)$ of the positron energies resulting form the process (3.7.6.-2) is of the form

$$F(\gamma_e) = \frac{1}{2}(A^2 - 1)^{-1/2} \int\limits_{\gamma^-}^{\gamma^+} F(\gamma_\pi)\,(\gamma_\pi^2 - 1)^{-1/2}\,d\gamma_\pi \tag{3.7.6.-5}$$

with $\gamma^\pm = A\,\gamma_e \pm (A^2 - 1)^{1/2}\,(\gamma_e^2 - 1)^{1/2}$

$$A = \frac{1}{4}\frac{m_\pi}{m_e} \tag{3.7.6.-6}$$

(RAMANATY and LINGENFELTER, 1966).

From the average number of positrons produced per proton the radio emission can be estimated if synchrotron radiation of positrons is assumed. The resulting spectrum of the synchrotron emission is then extended over more than five decades which is normally not in a best agreement with the radio observations even when the RAZIN effect is taken into account. Thus this mechanism is capable to account for only rare cases of exceptional big events.

A criterion for the applicability of the above cited mechanism can be found by observations of the gamma-radiation resulting as a consequence of the pair production with electrons. This radiation should be detectable below the COMPTON wave length.

3.8. Relations between the particle energy spectrum and the emitted radiation

3.8.1. Stationary conditions

Analogeously to the observed energy spectra of the non-solar cosmic radiation satellite measurements of solar particles after flares show an energy distribution for solar electrons approximately of the form

$$N(E)\,dE = K \cdot E^{-\gamma'}\,dE \qquad [\mathrm{cm}^{-3}] \,. \tag{3.8.1.-1}$$

If this distribution is inserted in the formulae given in section 3.5.2, the intensity of the synchrotron radiation is obtained as

$$I_\nu = \int_0^\infty \int_{E_0}^\infty \frac{W_\nu(E)}{4\,\pi} N(E, s)\,dE\,ds = \frac{\sqrt{3}\,L\,e^3\,K\,H}{4\,\pi\,m_e\,c^2} \int_{E_0}^\infty F\left(\frac{\nu}{\nu_c(E)}\right) E^{-\gamma'}\,dE\,, \tag{3.8.1.-2}$$

where I_ν is supposed as isotropic and reabsorption effects along the ray path s are neglected.

For $F(\nu/\nu_c)$ the approximation given by WALLIS (1959) can be used

$$F\left(\frac{\nu}{\nu_c}\right) = \int_{\nu/\nu_c}^\infty = K_{5/3}(z)\,dz = q\left(\frac{\nu}{\nu_c}\right)^{\alpha^*} \cdot e^{-\nu/\nu_c} \tag{3.8.1.-3}$$

with

$$q = 1.78 \quad \text{and} \quad \alpha^* = 1/3\,.$$

Hence it follows

$$I_\nu = \frac{q}{4\,\pi} L\,C\,K\,H\,\nu^{1/3}\,\varepsilon^{-1/3} \int_{E_0}^\infty E^{-\frac{\gamma'+2}{3}}\,c^{-\frac{\nu}{\varepsilon E^2}}\,dE\,, \tag{3.8.1.-4}$$

where

$$C = \frac{\sqrt{3}\,e^3}{m_e\,c^2}\,, \qquad \varepsilon = \frac{3}{4\pi} \cdot \frac{e\,H}{m_e^3\,c^5}\,.$$

With the substitution

$$x = E \cdot \sqrt{\frac{\varepsilon}{\nu}} \tag{3.8.1.-5}$$

one obtains

$$I_\nu = \frac{q}{4\,\pi} L\,C\,K\,H \left(\frac{\nu}{\varepsilon}\right)^{-\frac{\gamma'-1}{2}} \int_{x_0}^\infty x^{-\frac{\gamma'+2}{3}}\,e^{-\frac{1}{x^2}}\,dx\,. \tag{3.8.1.-6}$$

This denotes, that under the above made assumptions the intensity of the synchrotron radiation without reabsorption is proportional to $\nu^{-(\gamma'-1)/2}$. In the case of weak reabsorption it was already obtained by (3.5.4.-10) that no dependency from the exponent of the energy distribution function results.

Correspondingly, in more general cases the integral equation (3.8.1.-2) is to be solved, which under suitable conditions can be reduced e.g. to a LAPLACE transform.

3.8.2. Temporal changes of the energy distribution and the emitted spectrum

Variations of the energy distributions are due to temporal changes of the energy supply and losses (acceleration and deceleration processes) and to the escape or injection of particles with respect to a considered emitting volume. After dealing with the acceleration processes in the previous sections, now, especially the energy loss processes are of interest taking effect on fast particles (electrons) in a magnetoplasma. Such losses are represented by the following processes:

a) Gyro-synchrotron radiation
b) COULOMB bremsstrahlung (inelastic collisions)
c) Inverse COMPTON effect, i.e. scattering of fast electrons at photons in a radiation field
d) Generation of plasma waves
e) Ionization and excitation of atoms
f) Escaping of particles from the emitting region
g) Non-dissipative processes changing the energy distribution, e.g. elastic collisions, convection, heat transfer etc.

Estimations of the time variation of the energy distribution considering losses due to gyro-synchrotron radiation, collisions, and escaping of trapped particles have been made by TAKAKURA and KAI (1966).

The energy loss due to synchrotron radiation is

$$-\frac{dE}{dt} = W(E) \qquad (3.8.2.\text{-}1)$$

whereas the collisional losses are described by the so-called energy exchange time

$$t_e = \frac{E^2}{\langle (\Delta E)^2 \rangle} \qquad (3.8.2.\text{-}2)$$

defined by SPITZER (1962).

The escape of particles (electrons) from the considered volume is determined by

$$N_e(E, t)\, dE = N_e(E_0) \exp\left\{-\int_0^t t_D^{-1}\, dt\right\} dE_0 \qquad (3.8.2.\text{-}3)$$

whereas

$$t_D = \frac{v^2}{\langle(\Delta v_T)^2\rangle} \tag{3.8.2.-4}$$

is SPITZER's deflection time.

Assuming a power law distribution of the initial electron energies of the form (3.8.1.-1) and the action of the above cited three loss processes, the time dependency of the energy spectrum follows

$$N_e(E, t) = K\left(\frac{E}{m_e c^2}\right)^{-\gamma'} \left[\exp\left\{(1-\gamma')\frac{16\pi^2 e^2 \nu_H^2}{3 m_e c^3} t - \frac{1}{\tau_D}\int\limits_0^t \left(\frac{E}{m_e c^2}\right)^{-3/2} dt\right\}\right]$$
$$\cdot\left\{\frac{3.6 \cdot 10^{-22} N_e c}{16\pi^2 e^2 \nu_H^2}\left(\frac{E}{m_e c^2}\right)^{5/2}\left[1 - \exp\left(-\frac{5}{2}\cdot\frac{16\pi^2 e^2 \nu_H^2}{3 m_e c^2} t\right)\right] - 1\right\}^{-\left(\frac{3+2\gamma'}{5}\right)} \tag{3.8.2.-5}$$

$$(\tau_D \approx 2.4 \cdot 10^{12} N_e^{-1})$$

(TAKAKURA and KAI, 1966).

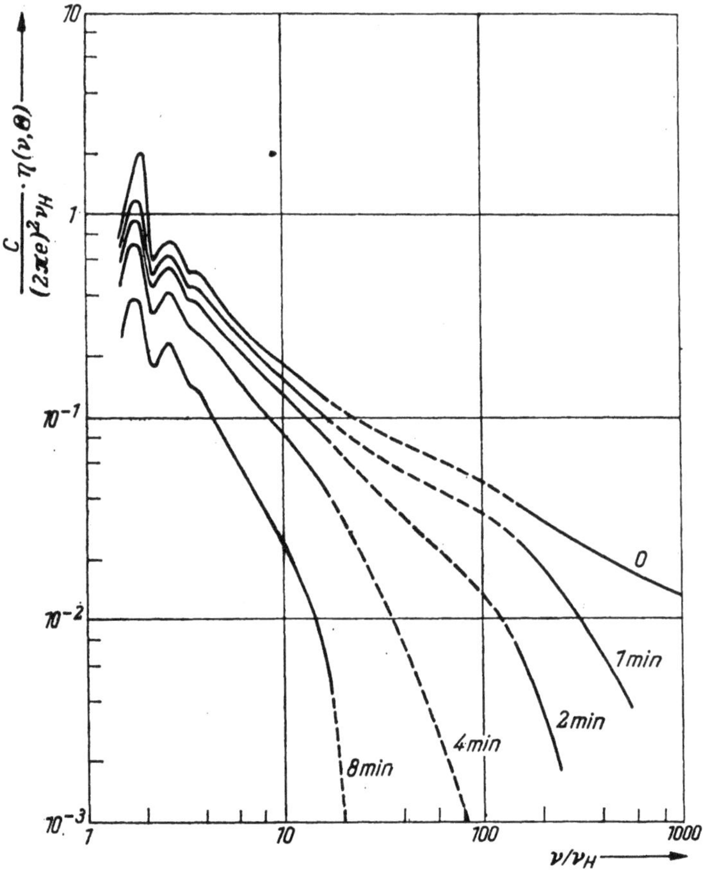

Fig. 23. Example for the time variation of the volume emissivity for gyro-synchrotron radiation computed for a burst model with $\gamma = 2$, $\theta = 60°$, $\nu_H = 1980$ MHz, and $N = 10^9$ cm^{-3} after TAKAKURA, UCHIDA and KAI (1968)

Basing on the knowledge of the energy distribution of the electrons the time variation of the emitted gyro-synchrotron spectrum has been calculated by TAKAKURA et al. (1968) for special cases. An example of their results is shown in fig. 23. It should be mentioned, that the peaks of the emission near the low harmonics of the gyrofrequency are flattened if inhomogeneous models of the burst source as occuring in reality are considered.

A further discussion of the time characteristics of the burst radiation basing on the equation of energy balance will be given in section 4.2.

4. The interpretation of radio bursts

The aim of this chapter is now to get a synthesis between the observational results and the theoretical fundamentals coming to a physical explanation of the us interesting features of solar radio bursts. For this purpose, however, it appears insufficient to consider only the radio bursts as isolated phenomena because evidently not so much an explanation of single burst characteristics, but the unambiguous interpretation of the whole phenomenon resulting from each of its structural details is needed. Therefore we try to understand the burst phenomena as a part of the total flare-burst complex comprising a multitude of different phenomena and physical processes.

We have to consider especially the time characteristics of the radio bursts and the burst spectrum flowing beyond others into the theory of the type IV bursts and proton flares including the emission and dynamical processes. Then we have to obtain model assumptions for the whole complex of compound bursts and to get conclusions for a possible forecast of solar events.

Looking at first at the time history of an solar burst phenomenon in general, three main stages of its development can be distinguished:

a) A storage of energy (pre-burst phase)
b) The sudden release of energy leading to a primary acceleration of particles (impulsive and explosive phases)
c) The expansion of the perturbation (involving secondary accelerations) and recovery.

These points now will be discussed in the following.

4.1. Storage and release of energy in connection with radio burst phenomena

4.1.1. Processes of energy storage

Starting with the basic question of the origin of the energy exchanged in the flare-burst complex we must confess, that similar as in the case of the acceleration mechanisms, our knowledge still is very limited. According to section 3.7 magnetic, mechanical, and nuclear energy have been considered. Without doubt the magnetic origin appears most probable to account for the flare-burst energy because in most cases a direct connection to nuclear energies seems not promising and the content of thermal or non-thermal kinetic energy appears insufficient to account for the large energy exchange in solar flare-burst events. Of course, here is a link to the more general question of the origin of solar activity at all, which is another big and unsolved problem of cosmical physics. However, the current ideas explaining the roots of the solar activity as a consequence of the action of convection, differential rotation, and subphotospheric magnetic fields appear to be capable to describe the phenomenon in its outlines. But further details of these questions must remain out of the scope of our present considerations.

As stated in 3.7.1, in principle the amount of magnetic energy stored in an active region would be sufficient for an explanation of the released energy but other difficulties must be tackled: At first, apparently, the locations of a maximal storage of magnetic energy are not identical with the location of maximal energy exchange. Maximal magnetic field strengths are present in photospheric and subphotospheric layers which seem to be untouched by the flare phenomenon. In this way not simply the whole content of magnetic energy represented by $H^2/8\pi$ must be considered as an effective potential source for the flare-burst energy, but only those parts of it, where by a favourable configuration an instable situation arises and the energy exchange becomes possible. The other difficulties arising from the suddeness of the dissipation process have been discussed already in 3.7.1. Informations about the role of the magnetic fields would be improved, if observations in various heights of the solar atmosphere during flare events become available. Nevertheless it seems clear that there are two extrema of magnetic field changes, namely dissipative and non-dissipative variations of magnetic fields, because large and long-term changes of magnetic fields occur without associated flare or burst events (cf. VYAL'SHIN et al., 1967). On the other side, the changes of magnetic fields connected with dissipative processes are rather small and observations of changes of one per cent or less would be required in short time intervals. One has to argue that the suddeness of the change of the magnetic field strength has something to do with the energy output leading to flare-burst events.

Unfortunately optical measurements which reveal the best angular resolution are scarcely possible at higher than photospheric levels, so that most of the informations have to come from radio observations with their worse resolution power.

4.1.2. Estimations of magnetic field strength from radio observations

Besides the attempts of an estimation of coronal magnetic fields from characteristics of drift bursts, e.g. the polarization of type I and type III bursts, the drift velocity and bandsplitting of type II bursts, etc. (ref. in TAKAKURA, 1966), also from the radio bursts of continuous radiation some informations about magnetic fields in the chromosphere and corona can be deduced. Especially the consideration of microwave bursts exhibits additional possibilities to obtain estimations of magnetic fields in their source regions.

The application of different methods of calculation of magnetic field strengths requires primarily the knowledge of the emission mechanism. Assuming bremsstrahlung, the magnetic field can be derived from the spectrum according to equation (3.3.5.-14) or from the degree of polarization. The latter becomes very simple in the case of very small optical depths where the degree of circular polarization depends only on the ratio of the gyrofrequency and observing frequency (3.3.5.-19). But at present it would be difficult to get conclusions about rapid temporal changes of the magnetic field — apart from spatial variations — during the life time of a burst event, as long as informations about the distributions of the essential parameters, e.g. density, temperature or particle energy and the dimension and structure of the emitting volume are not available. Thus, yet only temporal and spatial mean values of the magnetic field (as well as of the other parameters) can be expected in general.

Invoking synchrotron emission also different possibilities for an estimation of magnetic fields arise. For broad-band continua the position of the spectral maxima on the frequency axis gives an indication of the average magnetic field strength. According to TAKAKURA (1960) for bursts of medium energy without regard toresonance absorption effects the position of this maximum is at about $2\,\nu_H$. Taking into account the effects of resonance reabsorption, which become important for smaller particle energies, the position of the spectral maximum is roughly determined by $4\,\nu_H$ which leads to smaller magnetic field strengths than in the former case (GINSBURG and ZHELEZNYAKOV, 1959).

Concerning synchrotron radiation from relativistic particles without reabsorption for $\Phi = \frac{\pi}{2}$ and a monoenergetic distribution according to (3.5.3.-4) the magnetic field is given by

$$H = 0.71\, \nu_{\max} (1 - \beta^2) \qquad [\text{gauss}] \qquad (4.1.2.\text{-}1)$$

if $\nu_{\max}$ denotes the frequency of the spectral maximum. If the RAZIN effect becomes important, after (3.5.3.-9) it can be estimated

$$H_T \lesssim \frac{20\, N_e}{\nu_{\max}^2} \qquad (4.1.2.\text{-}2)$$

For the cutoffs of the synchrotron radiation spectrum at a frequency ν_g, in the RAZIN approximation $0 < 1 - n_j \ll 1$, $Y \ll 1$, an upper limit of the magnetic field in the cut-off region can be deduced

$$H < \frac{\nu_g}{2{,}8 \sqrt{1 - \beta^2}} \qquad (4.1.2.\text{-}3)$$

if the emission region has a sufficient well defined boundary in its height extension.

Another method of a derivation of magnetic fields is resulting from observations of the transitional frequency by the reversal of the sense of circular polarization. As discussed in 3.3.3 according to COHEN (1960) the magnetic field can be derived from the position of the transitional frequency in the critical region $Q = 1$ in a QT-region. The such obtained magnetic fields are not related to the source region of the burst emission and therefore about 1—2 orders of magnitude lower than the values derived by the other above mentioned methods.

In an alternative assumption by TAKAKURA (1960) the reversal of polarization is interpreted as an effect of resonance absorption (cf. 3.3.3). The hence resulting magnetic field strengths are typically in the range 500—700 gauss and by this also in good accordance with the results of other estimations.

Further methods are based on the calculation of the radiation losses by synchrotron emission depending on the magnetic field. A not widely known proposal for the estimation of the magnetic field strength from the decreasing rates of impulsive microwave bursts has been made by C. and F. CHIUDERI (1966). These authors use the assumption that after the time of the burst maximum no further energy supply takes place in the burst source and that the energy losses are only determined by synchrotron radiation.

Then from the differential equation for the time variation of the energy

$$-\frac{dW}{dt} = \frac{2\,e^4\,H_T^2}{3\,m_e^2\,c^3}\left[\left(\frac{E}{m_e\,c^2}\right)^2 - 1\right] = P(H_T, E, t) \qquad (4.1.2.\text{-}4)$$

and the substitution

$$x = \frac{1}{(1-\beta^2)^2} - 1 = \frac{E}{m_e\,c^2} - 1 \qquad (4.1.2.\text{-}5)$$

for a constant H_T the solution follows

$$\frac{x}{x+2} = \frac{x_0}{x_0+2}\,e^{-2yt}\,, \qquad (4.1.2.\text{-}6)$$

where

$$y = \frac{2\,e^4\,H_T^2}{3\,m_e^2\,c^5}\,. \qquad (4.1.2.\text{-}7)$$

x_0 denotes the initial value at the time $t = 0$.

The decrease of the radiation after the maximum is characterized by the half power time t_r which is defined as follows

$$P(H_T, E_0, t_r) = \frac{1}{2}\,P(H_T, E_0, 0)\,, \qquad (4.1.2.\text{-}8)$$

where $t = 0$ is the time of the burst maximum.

From (4.1.2.-6) and (4.1.2.-8) it follows

$$t_r(H_T, E_0) = \frac{1}{2\,y}\left[\ln 2 - 2\ln\frac{2\,x_r}{x_0}\right] \qquad (4.1.2.\text{-}9)$$

or

$$\bar{t}_r(H_T) = [\int_{x_1}^{x_2} N(x_0)\,dx_0]^{-1} \int t_r(H_T, x_0)\,N(x_0)\,dx_0\,, \qquad (4.1.2.\text{-}10)$$

where

$$N(x_0) = K\,x_0^{-\gamma'}\,. \qquad (4.1.2.\text{-}11)$$

C. and F. CHIUDERI (1966) showed that the influence of the quantities x_1, x_2, and x on t_r is neglectable with respect to the calculation of H_T and obtained finally

$$H_T = \left[\frac{3\, m_e^2\, c^5}{4\, e^4} \frac{\ln 2}{\bar{t}_r}\right]^{1/2} = 13.4 \cdot 10^3 \cdot (\bar{t}_r\,[\text{sec}])^{-1/2} \quad [\text{gauss}]\,. \qquad (4.1.2.\text{-}12)$$

The magnetic fields calculcated after this formula are typically above $1.5 \cdot 10^3$ gauss even for smaller events. As a consequence of the above made presumptions the obtained values can be considered as upper limits for the actual strength of the magnetic field in the burst volume.

A compilation of average values of the magnetic field obtained by different methods is shown in fig. 24.

4.1.3. The S-component as an indicator for the storage of kinetic and magnetic energy

Because we are still far from obtaining a satisfying picture about the magnetic field and its structure in several heights of the solar atmosphere, again some

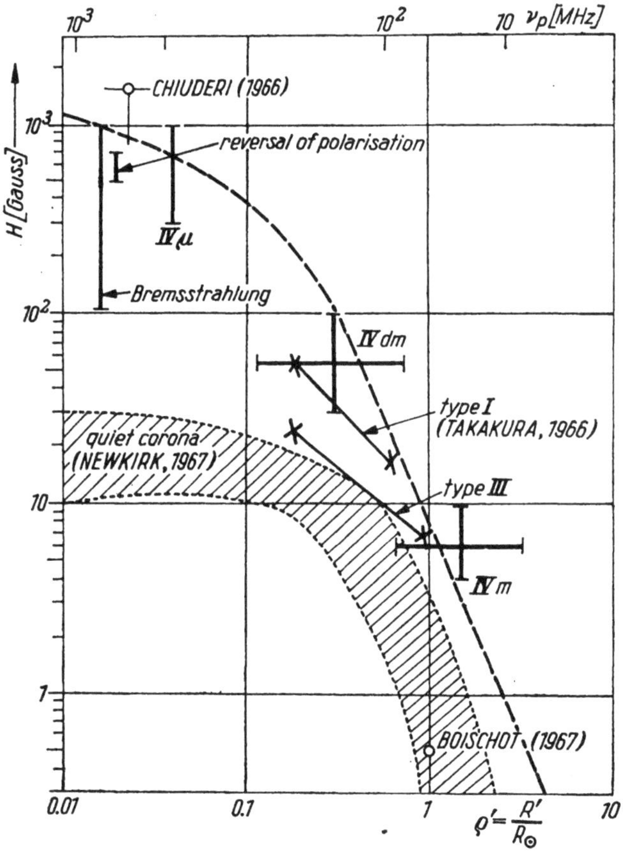

Fig. 24. Some results of estimations of coronal magnetic fields from radio observations

theoretical considerations should be added about that what we can expect concerning the energy supply in active regions.

At first we ask for the effect of magnetic fields on the normal transport of mechanical energy in the solar atmosphere. The influence of the magnetic field of active regions must result a hindering of the energy transport by particles perpendicular to the magnetic lines of force. If the diffusion of particles is considered, for which the velocity is coupled with a mechanical energy flow, a drift velocity is obtained by

$$\boldsymbol{W}_D = \boldsymbol{W}'_D + \boldsymbol{W}''_D , \tag{4.1.3.-1}$$

where the components $\boldsymbol{W}'_D$ and $\boldsymbol{W}''_D$ are caused by the curvature of the magnetic field lines:

$$\boldsymbol{W}'_D \sim m_e w_D^2 \frac{\boldsymbol{B} \times (\boldsymbol{B} \cdot \nabla) \boldsymbol{B}}{B^4} \tag{4.1.3.-2}$$

$$\boldsymbol{W}''_D \sim m_e w_T^2 \frac{\boldsymbol{B} \times \nabla \boldsymbol{B}^2}{B^4} \tag{4.1.3.-3}$$

(cf. SCHATZMANN, 1965).

For the escaping time t_D of relativistic particles of charge q from a volume with a characteristic length L the expression

$$t_D \approx \frac{q\, B\, L^2 (1 - \beta^2)^{1/2}}{2\, T_0} \tag{4.1.3.-4}$$

is valid (STEPANYAN and VLADIMIRSKY, 1961).

Hence, already in this relatively rough approximation, the essential dependency of t_D from the particle energy follows leading to the fact, that e.g. protons escape faster than electrons from a given plasma volume.

The resulting increase of the electron density in solar active regions beyond others leads to the development of the coronal condensations which are observed in the S-component of the radio emission. As an observational fact, there is a close connection between the emission of the S-component (not only in the radio region, but also in the X-ray region etc.) depending on the emission measure, the electron density, and the magnetic field and the occurrence of burst events.

Naturally the question arises, whether the quantity of kinetic energy stored in coronal condensations and accumulated for longer time intervals could be sufficient for an explanation of the flare phenomenon (CARMICHAEL, 1964, ELLIOT, 1964). As discussed in section 3.7.1. an accumulation of kinetic energy of the order of NkT $\sim 10^{32}$ erg in solar active regions would be required. Though this is more than the instantaneous energy content of both, the undisturbed chromosphere and corona ($5 \cdot 10^{31}$ erg), energies of about 10^{31} erg would be accumulated if the particles of the solar wind could be stored during the time interval of one day ($\sim 10^5$ s) over the area of an active region of about 10^{-3} of the solar disk. In this way, besides possible additional slowly acting dissipative processes of magnetic energy, the flare phenomenon has been regarded as a discontinuous release of the stored kinetic energy in active regions. But the untenibility of this imagination becomes evident as soon as e.g. the physical properties of the S-component are considered as it is shown in the following.

An estimation of the kinetic energy NkT under plausible assumptions about the size of the related volumes in the solar atmosphere leads to values of NT of the order 10^{18} grad cm^{-3} and more. This would result in an enormous heating of the active region which is not observed. If, on the other side, energies corresponding to NT $\sim 10^{16}$ grad cm^{-3} are taken as observed by the S-component, the needed characteristic dimensions of the storing region are too high by about 1—2 orders of magnitude. This discrepancy becomes still greater if it is demanded that only a relatively small fraction of the energy content of an active region is transformed by a flare event, because a breakdown of the coronal condensations after flares is not observed and the radiation of the S-component is usually not influenced even by largest events.

To overcome these difficulties, Elliot (1964 and 1969) improved these ideas by assuming an energy store by trapped protons which would be hardly detected by radiation. The protons again would be trapped in magnetic fields over a period of about 10^5 s built up an energy flux from below through the photosphere of approximately 10^8 erg cm^{-2} s^{-1}. According to Elliot (1969) the release of a fraction of these stored particles into lower levels would produce an upward ejection of photospheric matter into the trapping region triggering off a rapid degradation of the protons there together with a further dumping into the lower atmospheric levels.

But it remains very questionable for the reason mentioned above, that protons of high energies can be trapped for periods of the order 10^5 s in solar magnetic fields. Furthermore it must be noted, that the details of the flare process are not yet clear and quantitative calculations of the mechanism are yet missing.

Thus we are forced to come back to the old conclusion of considering the magnetic energy as the at present most probable direct source of the flare-burst energy. During the development of an active region the largest magnetic fields are present in the E- and F-stage of the sunspots. In these phases usually also the most and largest flare-burst events occur. The S-component exhibits in these stages typically a spectral maximum in the range $\lambda = 5\cdots10$ cm demonstrating the influence of. magnetic fields in chromospheric and coronal heights. If a greater angular resolution would be become available it is expected that the observation of structural details will provide a further clue on the critical points of the flare and burst origin. Besides the radio emission also the slowly varying X-ray emission indicates enhanced particle densities especially at the maximum stages of the development of an active region. In this way the spectra of the slowly varying X- and radio emission reflect well the global content of both, magnetic and kinetic energy.

4.2. The time characteristics of solar radio bursts

4.2.1. Basic factors determining the burst morphology

Now we consider the dynamical processes which determine the time characteristics of the burst radiation. After that, what was said in the previous chapters it must be clear, that we are only in the beginnings of an full understanding of theses processes. Nevertheless, we try here to discuss some basic ideas and attempts of an approach to the present problems.

The time profile of the flux density of a continuous radio burst as observed at a certain frequency is determined in a first order by the time dependent (and space dependent) energy distribution of the radiating electrons. Temporal changes of the energy distribution are expressed by an equation of energy balance, which may be given in the following form

$$\frac{\partial N(E,t)}{\partial t} + \left[\sum_{\varkappa=0}^{2} (A_\varkappa - B_\varkappa) E^\varkappa\right] \frac{\partial N(E,t)}{\partial E} + \left[\sum_{\varkappa=0}^{2} (A_\varkappa - B_\varkappa) E^{\varkappa-1}\right] N(E,t) = Q_1(E,t) - Q_2(E,t) \,. \tag{4.2.1.-1}$$

This equation is a special form of the well known general equation

$$\frac{\partial N(E,t)}{\partial t} + \frac{\partial}{\partial E}\left(N(E,t)\frac{dE}{dt}\right) = Q(E,t) \tag{4.2.1.-2}$$

which has been applied for problems of extragalactic radio astronomy e.g. by GINSBURG (1958), KELLERMANN (1964, 1966), HIRTH (1967) and others.

In (4.2.1.-1) it was put

$$\frac{dE}{dt} = \sum_{\varkappa=0}^{2} (A_\varkappa - B_\varkappa) E^\varkappa \,, \tag{4.2.1.-3}$$

where acceleration and energy loss processes have been described by the coefficients $A_\varkappa$ and $B_\varkappa$, respectively. In particular we consider

A_0, B_0 — recombination and ionization processes
A_1 — FERMI acceleration
B_1 — bremsstrahlung
A_2 — COMPTON effect and synchrotron acceleration
B_2 — synchrotron radiation, inverse COMPTON effect.

The Q-terms represent the energy supply from outward, in particular Q_1 and Q_2 describe the injection and ejection of particles, respectively.

In the following now the time characteristics of some special morphological burst types will be discussed by means of the equation of energy balance.

4.2.2. Quasi-stationary processes — gradual bursts

At first the gradual bursts occurring in the microwave region will be considered, for which the equation of balance may be written in the following form

$$\frac{\partial N(E, t)}{\partial t} + C_1 E \frac{\partial N(E, t)}{\partial E} + C_1 N(E, t) = 0 . \qquad (4.2.2.\text{-}1)$$

Here the abbreviation

$$C_\varkappa = A_\varkappa - B_\varkappa \qquad (4.2.2.\text{-}2)$$

was introduced.

The time characteristics can be characterized schematically by the following phases

$$\left.\begin{array}{ll} \text{rise} & t_1 \leqq t < t_2\colon C_1 > 0 \\ \text{maximum} & t_2 \leqq t \leqq t_3\colon C_1 = 0 \\ \text{fall} & t_3 < t \leqq t_4\colon C_1 < 0 . \end{array}\right\} \qquad (4.2.2.\text{-}3)$$

For the stage of the burst maximum it can be assumed

$$\frac{\partial N(E, t)}{\partial t} = 0 \qquad (4.2.2.\text{-}4)$$

and the case of a quasi-stationary process is present. The energy distribution remains constant without necessity of a determination of the initial conditions.

If $C_1 = 0$, retaining the condition of quasi-stationarity there follow

$$C_1 E \frac{\partial N(E, t)}{\partial E} = - C_1 N(E, t) \qquad (4.2.2.\text{-}5)$$

and

$$N(E, t) = \text{const}\, E^{-1} . \qquad (4.2.2.\text{-}6)$$

4.2.3. Impulsive processes and the decreasing phase of the burst emission

Next we consider weak impulsive microwave bursts. In this case we propose tentatively a predominance of bremsstrahlung against synchrotron radiation. Then the equation of balance becomes

$$\frac{\partial N(E, t)}{\partial t} + C_1 E \frac{\partial N(E, t)}{\partial E} + C_1 N(E, t) = Q_1(E, t) - Q_2(E, t) = Q(E, t) . \qquad (4.2.3.\text{-}1)$$

The time characteristics again can be divided into

$$\left.\begin{array}{ll} \text{increase} & t_1 \leqq t < t_2 : C_1 > 0,\ Q = a\, E^{-g} \\ \text{maximum} & t = t_2 : C_1 = 0,\ Q = 0 \\ \text{decrease} & t_2 < t \leqq t_3 : C_1 < 0,\ Q = 0 . \end{array}\right\} \qquad (4.2.3.\text{-}2)$$

Of course, also other assumptions, e.g. for the Q-term like $Q_1 = a\, E^{-g}\, e^{-b/E}$ or the introduction of corresponding expressions for Q_2 etc. could be made.

Special solutions for different initial conditions can be obtained, e.g.

$$N(E, t_2) = a\, E^{-g}$$
$$N(E, t) = a\, E^{-g}\, e^{C_1(g-1)(t-t_2)} \,. \tag{4.2.3.-3}$$

Now the more important case of stronger impulsive bursts will be considered, where synchrotron radiation is assumed to be predominant. Instead of (4.2.3.-1) it is written

$$\frac{\partial N(E, t)}{\partial t} = C_2{}'\, E^2 \frac{\partial N(E, t)}{\partial E} + C_2\, E\, N(E, t) = Q(E, t) \tag{4.2.3.-4}$$

with the following stages

$$\left.\begin{array}{ll} \text{increase} & t_1 \leqq t < t_2 : C_2 > 0,\ Q \neq 0 \\ \text{maximum} & t = t_2 : C_2 = 0,\ Q = 0 \\ \text{decrease} & t_2 < t \leqq t_3 : C_2 < 0,\ Q = 0\,. \end{array}\right\} \tag{4.2.3.-5}$$

For a study of the declining phase as "initial" value again the following expression may be arbitrarily taken

$$N(E, t_2) = a\, E^{-g}\,. \tag{4.2.3.-6}$$

Then the solution follows

$$N(E, t) = a\, E^{-g}\,(1 - C_2\, E\, t)^{g-2} \tag{4.2.3.-7}$$

which describes an energy spectrum changing with time and tending to zero for large t. The solution of the inhomogeneous differential equation (4.2.3.-4) with

$$Q(E, t) = K\, E^{-\gamma'} \tag{4.2.3.-8}$$

is given by

$$N(E, t) = \frac{K}{C_2\,(\gamma' - 1)}\, E^{-(\gamma'+1)}\,[1 - (1 - C_2\, E\, t)^{\gamma'-1}] + a\, E^{-g}\,(1 - C_2\, E\, t)^{g-2}\,. \tag{4.2.3.-9}$$

The resulting energy spectrum consists of two different parts. For large E the influence of the injection spectrum is dominating, while for smaller values of E the influence of the initial condition becomes important. The difference between the solutions (4.2.3.-7) and (4.2.3.-9) can be used for a test of the action of injection mechanisms. Comparisons with real burst events have shown that in many cases the injection of particles does not play an important role during the post-maximum phase of the development of simple radio bursts.

It must be mentioned that the above given solutions possess a physical meaning only till the point $C_2\, E\, t = 1$, otherwise the expressions given in parentheses become negative. The reason for this strange behaviour lies in the arbitrary assumption (4.2.3.-6) which, strictly speaking, contradicts the concept of the energy balance if the whole range $0 \leqq E \leqq \infty$ is considered. This difficulty can be overcome by other assumptions following a discussion of the general solution of the differential equation (4.2.1.-1) with precise initial conditions (non-monotonic distributions for $N(E, t_0)$, which can be obtained e.g. by a LAPLACE transform.

4.2.4. Burst groups and the phenomenon of homologous bursts

Sometimes a conspicuous repeating of burst events and in rare cases also a resemblance in special morphological details can be observed for bursts occurring in a certain spectral range within more or less short time periods, notwithstanding the fact that actually each burst event must be regarded as an individuum. Characteristically such groups originate in special active regions. Burst groups are defined as a series of events if the time interval between the different bursts are only a few multiples of their duration.

Analogeously to the phenomenon of homologous flares (ELLISON et al., 1960) a concept of homologous radio bursts was introduced for sequences of bursts resembling in special characteristics (FOKKER, 1967; KRÜGER, 1967). Later homologeous X-ray bursts have been described (PINTER, 1969) and homologous bursts have been found also homologous optically (RAO, 1970).

For bursts groups and especially for homologous bursts interesting questions arise with respect to the maintenance of special structures in solar active regions. In comparison to the usually applied classification scheme of morphological burst types the investigation of homologous burst structures implies an advanced stage of looking into recurrencies of solar phenomena. But defining homologous structures first of all it must be stated, that observations of spatial structural details as essential for the definition of homologous flares in the optical region, at present are not available in a sufficient number and quantity for the radio bursts. Therefore other characteristics of the solar burst emission have been treated which perhaps could give a criterion for the existence of homologous structures of the source region. Such characteristics are e.g.

a) The burst morphology (time profile) at single observing frequencies including maximal intensities
b) Slope and form of the spectrum of flux density
c) The frequency extent of the burst radiation
d) Polarization characteristics
e) Association with other radio burst components of different types
f) Association with homologous flares
g) Association with (homologous ?) X-ray bursts and other related phenomena.

Statistical investigations have shown that similarities in all above mentioned points are scarcely to be expected. Therefore, in practice, restrictions are necessary, whereas similarities concerning at best the points a)—c) have been used for a provisional definition of homologous bursts. But even homologous events of this category are not often observable (a few per year). Nevertheless these cases are of special interest for a study of the development of active regions.

Interpreting homologous bursts it must be taken into mind, that the centres of solar activity have a three-dimensional structure. If homologous bursts appear in a very limited part of the spectrum, e.g. only at special centimetric wavelengths, the conditions responsible for homology must be expected to be fulfilled in a relatively small part of an active region. For a larger frequency extent of homologous burst structures it seems plausible to assume that especially vertical structures in a greater range of heights in the solar atmosphere have been maintained during the observed time interval. In general, conditions which are favourable for the occurrence of homologous bursts appear to be realized easier in smaller than larger volumes. From the time interval of the occurence of homologous events which is typically $\leqq 1 - 2$ days an impression of the time scale of the change of single structures in solar active regions is obtained.

4.3. The burst spectrum

4.3.1. Monotonic spectra

As next we are dealing with the spectrum of the radio bursts. At first monotonic spectra will be considered, whereas the definition of "monotony" may be applied here to only a more or less restricted range of observed frequencies. (Monotonic burst spectra extending over the whole electromagnetic spectrum are, of course, not possible.)

Considering the microwave region, an important case is established by spectra rising monotoneously towards higher frequencies. In principle, those spectra can be produced either by bremsstrahlung or by synchrotron radiation of sufficient high particle energies. So far as bremsstrahlung is concerned according to 3.2.3 for the limiting case of homogeneous burst models and large optical thickness spectra of the form $I_\nu \sim \nu^\alpha$ with $\alpha = -2$ are expected, while the influence of small optical depths would be expressed by values $\alpha > -2$. A similar influence on the slope of the spectrum would be caused by inhomogeneous models of the source region. Spectra with slopes steeper than $\alpha = -2$ are here not expected, otherwise effects near the local plasma frequency must be inferred. The observed spectral range is determined by the particle number (volume), the particle energy (temperature), and the particle density in the emitting region as well as by the sensitivity of the used receiver.

For the gyro-synchrotron mechanism more or less extended emission bands are characteristical. Usually the spectral maximum of these bands is within the observed radio range so that non-monotonic spectra are received. A spectrum rising over several decades towards higher frequencies can be expected for electron energies of more than 1 MeV if the Razin effect is negligible. This would correspond to equivalent radiation temperatures of more than 10^{11} °K which refers to large type IV bursts. It should be noted, that in this case the slope of the spectrum can be steeper than in the case of bremsstrahlung. For real conditions, however, the distribution function of the particle energies and their spatial distribution as well as that of the magnetic field play an important role. An unique estimation of the physical parameters of the source region from the spectral characteristics of the burst emission is therefore only possible if special model assumptions are introduced.

4.3.2. Cut-off and cross-over frequencies

In contrast to the bremsstrahlung where the decrease of the spectrum down under the limit of detectability on the low frequency side does not occur very abruptly (so far as propagation effects are not to be taken into account), the gyro-synchrotron radiation has a low-frequency cutoff at the gyrofrequency which according to (3.4.1.-6) for relativistic electron energies is given by

$$\nu_g = \frac{1}{2\pi}\frac{e H}{m_e c}\frac{(1-\beta^2)^{1/2}}{1 - n_j \beta \cos\theta \cos\Phi} = \frac{2.8 \cdot 10^6 H_{[\mathrm{gauss}]}}{\gamma \left[1 - n_j\left(1 - \frac{1}{\gamma^2}\right)^{1/2} \cos\theta\cos\Phi\right]} \text{ [MHz]} \qquad (4.3.2.\text{-}1)$$

where

$$\gamma = \frac{E}{m_e c^2} = (1 - \beta^2)^{-1/2} . \qquad (4.3.2.\text{-}2)$$

In practice the position and sharpness of the low frequency edge of the spectrum depends essentially on the distributions of the particle energies, the pitch angles, and the magnetic field, if the influence of the medium by the Razin effect or cut-offs at the local plasma level (which would depend on the ratio X/Y) is negligible (cf. below). A distribution of the magnetic field strength over a large range as well as high particle energies result in an increasing frequency extent of the emitted radiation.

Hence for the occurrence of emission bands of relatively small frequency extent, as it is observed e.g. for many impulsive microwave bursts, it follows, that besides not too high particle energies also a spatial limitation of the source region (magnetic field!) must be required.

Further aspects are taken from the consideration of the influence of the ambient plasma medium, if the refractive index departs noticeably from unity. As discussed in section 3.5.3 the Razin effect causes an increase of cut-off frequency and a diminution of the emitted frequency range. Nevertheless a limited extension of the source region must be assumed, otherwise by the spatial distribution of the physical parameters a broadening of the spectrum would result. Similar conclusions about a restriction of the size of the emitting volumes especially in vertical direction must be drawn, if the low-frequency cut-off is caused by zeros of the refractive index.

The cross-over frequency in the medium decimetre region marking the boundary between the typical microwave burst emission and the metre region can be interpreted by the assumption that the magnetic field responsible for the gyro-synchrotron emission of the microwave bursts are typically between, say, 10^2 and 10^3 gauss, but not below this value. An emission at lower frequencies then is excluded, also for bremsstrahlung, because this would require unlikely high equivalent temperatures (cf. fig. 10).

In fig. 25 the position and extent of the spectrum of the gyro-synchrotron radiation in the one-particle approximation ($n = 1$, $\Phi = \pi/2$) are sketched by drawing the cut-off (cross-over) frequencies and high frequency boundary of the emission. Herewith a first rough impression of the emitted spectral range in dependence of particle energy and magnetic field on the basis of the single particle process is obtained. From the existence of the cross-over frequency one can conclude that the electron energies for bursts occurring in the microwave range do not exceed 10^6 eV. From fig. 25 one also gets an idea about the expected

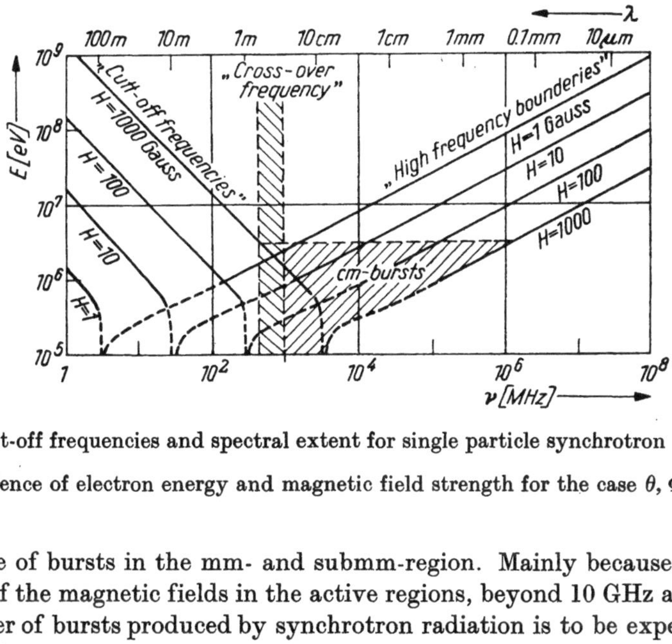

Fig. 25. Cut-off frequencies and spectral extent for single particle synchrotron radiation in dependence of electron energy and magnetic field strength for the case $\theta, \Phi = \dfrac{\pi}{2}$

occurrence of bursts in the mm- and submm-region. Mainly because of the restriction of the magnetic fields in the active regions, beyond 10 GHz a diminishing number of bursts produced by synchrotron radiation is to be expected. Assuming 10^4 gauss as an upper limit of solar magnetic fields, e.g. at 0.3 mm wavelength only bursts with electron energies greater than 10^6 eV can be observed.

For a quick comparison the connection between electron energy, Lorentz factor, and equivalent temperature for synchrotron radiation is represented in fig. 26.

4.3.3. Relation between frequency extent and position of the spectral maximum for synchrotron radiation

For relativistic electron energies according to (3.5.3.-4) under assumption of the single particle approximation and $n = 1$ the position of the spectral maximum is given by

$$\nu_{\max} \approx b \cdot \nu_c \,, \tag{4.3.3.-1}$$

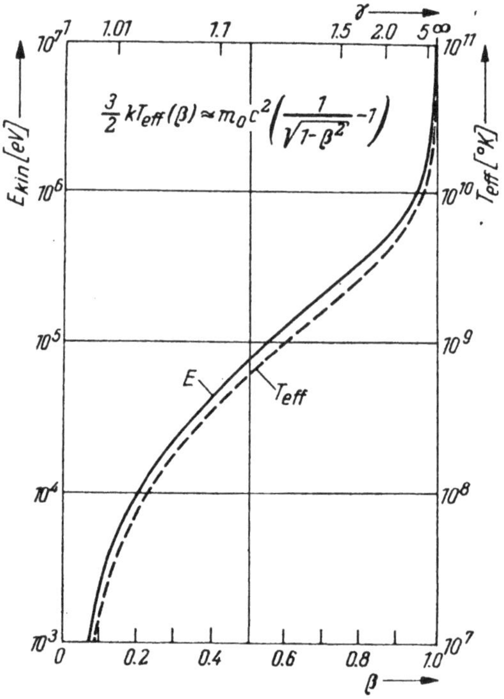

Fig. 26. Relation between electron energy E_{kin}, β, LORENTZ factor γ, and effective temperature T_{eff} for sychrotron radiation

where

$$\left. \begin{aligned} \nu_c &= 4.2\, H_{T\,[\mathrm{gauss}]} \left(\frac{E}{m_e c^2} \right)^2 [\mathrm{MHz}] \,. \\ b &\approx 0.3 \end{aligned} \right\} \qquad (4.3.3.\text{-}2)$$

Similarily also the connection between $\nu_{\max}$ and the frequency extent can be estimated for special values of the parameters H, Φ, and θ. Then the frequency extent of the burst continua for $n = 1$ is determined essentially by the specification of the average particle energy or its distribution function. The particle number and the size of the emission volume which influence the intensity of the burst radiation are here not of a great importance for this consideration, provided that the burst emission is strong enough to exceed the limit of detectability. Hence it is easily understood that statistical investigations do not show any clear connection between the maximal intensity and the frequency extent of the burst emission (cf. HACHENBERG and KRÜGER, 1964).

Another picture is obtained, if the relation between $\nu_{\max}$ and the frequency extent is considered. Here theoretically a connection is expected which indeed has been established by observations. In a wide range of the spectrum $\nu_{\max}$ and the the frequency of the high frequency boundary ν_{end} of the relativistic synchrotron spectrum esxhibit a similar dependency of the magnetic field strength so that the quotient of both frequencies becomes independent of the magnetic field.

It sould be emphasized once more that the above made considerations are strictly valid only for monoenergetic distributions of the electron energies within the framework of the one-particle approximation. The existence of an energy distribution depart from the δ-function must slightly modify our considerations. Furthermore it must be taken into account that for electron energies below 10^6 eV the approximation of the synchrotron radiation commonly used and given in 3.5.2. eads to considerable errors, and approaching regions near the plasma frequency[1], the influence of the medium remarkably diminishes the frequency extent. These reasons may be responsible for the observational fact that the ratio between the maximum frequency and the band width of the emissions of the cm-bursts reveals more a quadratic than a linear relation.

4.4. Theoretical aspects of the type IV bursts

4.4.1. Type IVμ bursts

As emission mechanism of the type IV μ bursts generally gyro-synchrotron radiation is accepted. In this respect, no, or merely quantitative differences are present between the type IV bursts and other burst types generated by synchrotron emission as impulsive or complex bursts in the microwave region.

Thus the distinction between type IV μ bursts and other burst types in the microwave region is based on the morphological properties of the time profile of the bursts (cf. 2.2.4). It must be noted, that the transitions between such different types appear fluently and are far from being sharp. Nevertheless it can be argued that the problem of the generation of different morphological burst types including the type IV burst involves the question after the spatial configuration and dynamical processes inside the emitting region, whereas the special mechanisms of triggering and particle accelereation take a special effect. The emission mechanism (i.e. in most cases gyro-synchrotron emission) then plays an unspecific role for the classification of bursts especially in the microwave region. However, at first we will disregard from the difficult questions of the dynamical process and consider other aspects of the theory of type IV bursts.

Some basic characteristics of the solar type IV μ bursts as the spectrum and the state of polarization can be explained already by a homogeneous burst model assuming a plasma layer of constant electron density and constant magnetic field strength. In order to explain effective temperatures of the order of 10^9 °K, electrons with speeds of $\beta = 0.6$, i.e. 10^5 eV would be sufficient provided that the optical depth is large enough:

$$\tau_\nu = \varkappa_{\nu,js} N_e L \gtrsim 1 . \qquad (4.4.1.\text{-}1)$$

This condition is fulfilled e.g. for $N_e \gtrsim 10^3$ cm^{-3} and $L \gtrsim 10^9$ cm. The magnetic field can be regarded as varying between values of the order of $2 \cdot 10^2$ and

$2 \cdot 10^3$ gauss. But taking the above mentioned low electron energies, the resulting frequency extent of the emitted radiation is somewhat small. For an interpretation of the actually observed bandwidths of several hundreds MHz therefore either inhomogeneous models with locally varying magnetic fields or higher particle energies ($\beta \gtrsim 0.8$) must be inferred. In the latter case one is not dependent upon the condition (4.4.1.-1) and smaller optical depths are also possible. If $\nu_H < \nu_p$, the influence of the medium on the average spectrum is not, or only weakly, expressed. For the interpretation of the weaker microwave bursts and not too large type IV-bursts in general the RAZIN effect is not necessarily to be taken into account. For strong type IV events being associated with the expellation of energetic particles, however, the existence of strong magnetic fields must be supposed and an effect of the medium on the frequency extent cannot be excluded. Cases with a characteristic decrease of the spectrum in the lower cm-region might be explained by the RAZIN effect.

The sense of the circular polarization which usually corresponds to the extraordinary wave mode is easily explained because just this property must be expected for synchrotron radiation at higher harmonics. The observed relatively low degrees of polarization ($\sim$50%) could indicate a double or multiple structure of the burst source corresponding to the distribution of the magnetic field polarities of complex spot groups.

Further possibilities of the interpretation of a low degree of polarization and its small variability with frequency are given e.g. by the assumption of inhomogeneous, vertical stratified models of the emission volumen. If, for simplicity, only the magnetic field is assumed as variable, then the resulting spectrum is composed of the sum of a multitude of frequency shifted spectral curves of similar shape. When gyro-resonance absorption effects by thermal electrons are taken into consideration, according to 3.4.2 the effect for the extraordinary component is by two or three orders of magnitude stronger than for the ordinary component. This causes a relative increase of the ordinary component at lower frequencies which is expressed by a decrease of the degree of polarization of the total emission. For a quantitative calculation of the degree of polarization after 3.5.1 also the form of the electron energy distribution must be known.

A dropping of the degree of polarization finally could be caused theoretically also by a contribution of linear polarized components. Such components would be recorded as a non-polarized emission due to the FARADAY effect along different ray paths and to the finite band width of the receiving equipments.

4.4.2. Type IVdm bursts

For the type IV dm bursts the main difficulties of a physical explanation arise from the relatively small frequency extent combined with sometimes very high tlux densities and the typically ordinary sense of circular polarization (according fo the leading spot hypothesis) whereas the degree of polarization amounts high values up to about 90 per cent (cf. 2.2.4.2). Furthermore the tendency to long durations and the relation to the storm continua in the metre region are somewhat puzzling.

In principle, a reversal from extraordinary into ordinary polarized waves can be explained by propagation effects. Such a mode transformation is expected theoretically at inhomogeneities in QT-regions as discussed in 3.3.3. Another possibility of such a transformation could be realized in regions near the gyrofrequency where for a radiation coming from regions where $Y > 1$ (Mollwo, 1970). But a proof of the actual role of different propagation processes in the solar atmosphere is not yet obtained.

Next the emission processes themselves must be considered. Though originally the gyro-synchrotron radiation has been proposed as emission mechanism for type IV bursts in general, the variety of existing type IV burst subtypes with different characteristics and especially the occurrence of ordinary polarized radiation components demands a deeper reconsideration of the question of the emission mechanisms.

Concerning the generation of a predominance of the ordinary wave mode on the basis of gyro-synchrotron emission two conditions can be considered. Firstly for arbitrary radio frequencies the parameters β, θ, and Φ must have such special values that the amount of the ordinary component exceeds that of the extraordinary one. In particular the condition

$$\gamma_{s=1} > 0.5 \qquad (4.4.2.\text{-}1)$$

must be fulfilled (cf. (3.5.1.-6)).

As it was shown by Böhme (1964) and Kai (1965), electrons with energies $\beta \geqq 0.9$ and very high pitch angles can explain a dominating polarization in the ordinary sense and high intensities, which follows immediately from the equations (3.4.1.-6) and (3.5.1.-6). But it appears questionable that electrons with such high energies could be trapped in the magnetic field for a period sufficient to explain the observed life time of the burst radiation. Moreover, it remains difficult to understand the observed small frequency extent of the type IV dm burst components. Therefore an explanation by high energy particles appears very doubtful.

A second possibility for a generation of a dominating polarization in the sense of the ordinary mode is represented by the condition

$$\nu_g(\beta, H, \theta, \Phi) < \nu < \nu_H\,, \qquad (4.4.2.\text{-}2)$$

where

$$\frac{\nu_g(\beta, H, \theta, \Phi)}{\nu_H} < 1$$

which is due to the strong absorption of the extraordinary mode by thermal electrons in the region $\nu < \nu_H$. An explanation of the type IVdm bursts has been proposed by Yip (1967) on the basis of this condition. However, this effect is restricted to low harmonics of the gyrofrequency (cf. 3.4.2). Therefore, taking into account the sometimes high intensities, also this possibility must be rejected to explain the polarization characteristics over the whole frequency range even if inhomogeneous models of the source region are proposed.

Avoiding these difficulties TAKAKURA and KAI (1961) and KAI (1965) invoked negative reabsorption of gyro-synchrotron radiation of weakly relativistic electrons. For the realization of this effect an anisotropic distribution of the electron orbits is required, which in application to type IV dm bursts has to support the generation of the ordinary component. Under these circumstances the following condition

$$\left.\begin{array}{l} N_e^{(\mathrm{syn})}\, \varkappa_{\nu,2s}^{(\mathrm{syn})}(\nu,\theta) + N_e^{(\mathrm{th})}\, \varkappa_{\nu,2s}^{(\mathrm{th})}(\nu,\theta) < 0 \\ N_e^{(\mathrm{syn})}\, \varkappa_{\nu,1s}^{(\mathrm{syn})}(\nu,\theta) + N_e^{(\mathrm{th})}\, \varkappa_{\nu,1s}^{(\mathrm{th})}(\nu,\theta) > 0 \end{array}\right\} \qquad (4.4.2.\text{-}3)$$

i.e.

$$\left[\frac{\varkappa_{\nu,2s}(\nu,\theta)}{\varkappa_{\nu,1s}(\nu,\theta)}\right]^{(\mathrm{th})} < \frac{N_e^{(\mathrm{syn})}}{N_e^{(\mathrm{th})}} \cdot \frac{\varkappa_{\nu,2s}^{(\mathrm{syn})}(\nu,\theta)}{\varkappa_{\nu,1s}^{(\mathrm{th})}(\nu,\theta)} < \left[\frac{|\varkappa_{\nu,2s}(\nu,\theta)|}{|\varkappa_{\nu,1s}(\nu,\theta)|}\right]^{(\mathrm{syn})}$$

must be realized.

This denotes a strong non-equilibrium distribution and a restriction of the number of emitting electrons and the range of the angle θ.

The concept of the amplification of waves by negative reabsorption up to now does not conflict with the observations. But further investigations are needed in order to verify the above made statements.

4.4.3. *"Moving" and "quasi-stationary" type IV bursts in the metre region*

Stimulated by polarization measurements at present the opinion is disseminated that the moving (IVmA) and stationary (IVmB, see next section) components of the type IVm bursts do not differ only by the state of motion of the source itself but also by the emission mechanism. Because of the extraordinary sense of the polarized radiation (according to the leading spot hypothesis) an interpretation by synchrotron emission has been proposed for type IVmA bursts. Analyzing the spectral features TAKAKURA and KAI (1961) estimated electron energies of about $1.3 \cdot 10^5$ eV and magnetic field strengths of 20 to 45 gauss as characteristical for the source regions of this component.

Thus not the emission, but the dynamics of the burst source makes difficulties. Up to now theoretical attempts for a treatment of the motion of the burst source are not satisfying developed. But also the observational material is still incomplete. It must be clarified to what extent a more or less well expressed state of motion of the burst source is typical for all type IVmA events. Up to present greater motions have been revealed only for a very limited number of type IV bursts. (WEISS, 1963; BOISCHOT and CLAVELIER, 1968). The motions of the order of 10^3 km/s are typically lower than the ALFVÉN velocity of the medium. The latter corresponds to the speed of the expansion of the flare region up to areas of 10^5 km diameter during the flash phase. The significance of a frequency drift as characteristical for moving components is not yet clearly established and the whole dynamical process not yet well understood. It must be mentioned that also reversals of the direction of the motion in later stages of the burst development have been observed and a correlation with geophysical effects is not very pronounced. With this respect the importance of the moving type IV bursts primarily would be greater for solar physics than for the aspect of solar-terrestrial physics.

With regard to the restricted number of clearly established moving components it must be supposed that obviously only a certain fraction of type IVmA bursts (generating synchrotron emission) exhibits a well defined movement of the burst source and even for "moving" type IVmA bursts the state of motion is not always extended over the whole burst duration. On the other side, non-moving type IVmA burst components or phases designated as "quasi-stationary" type IVmA bursts are typically expressed during the explosive phase and well correlated with terrestrial effects. Therefore a distinction can be made between (quasi-stationary) type IVmA$_1$ and (moving) type IVmA$_2$ bursts (Krüger, 1968; Böhme et al., 1970).

4.4.4. Stationary type IV bursts and continuum storms

For an interpretation of the so-called stationary or type IVmB bursts especially the high degree of circular polarization in the ordinary sense and the relatively small source diameters are of interest. Furthermore the greater duration and the gradual transition into continuum storms challenge our attention. Sometimes a slow drift of the onset from higher to lower frequencies and the typical increase of the degree of polarization during the life time of the event requires an explanation. Also the strong directivity and a certain indication for an origin near critical frequencies must be interpreted (cf. 2.2.4.3).

The above mentioned characteristics contradict an explanation by simple (electromagnetic) synchrotron radiation, though Takakura (1962) hypothetically proposed such a radiation from the following spot region. Furthermore, because of their characteristics (excluding the sense of polarization), type IVmB and type IVdm bursts must be considered as physically distinct phenomena.

With respect to the time variation, in principle, two alternative ideas can be considered. One idea consists in the assumption of the existence of the action of at least two different emission mechanisms relieving one another during the life time of the burst. In this case the reasons for the occurrence of such a sequence in time must be cleared up.

The second general possibility of an explanation of the time characteristics of the type IVmB bursts suggests the action of one and the same emission mechanism under a temporal variable configuration of the source region which would be mainly related to local magnetic fields. Kai (1965) assumed an originally regular structure of the magnetic field which becomes more and more irregular by the impact of particles. Subsequently a stable configuration is to be built up during a phase of recovery emitting radiation preponderantly in the ordinary component. As relaxation time for this process it is expected

$$\tau \sim \frac{L}{v_m} \tag{4.4.4.-1}$$

which for $N \approx 5 \cdot 10^8$ cm^{-3} and $L \approx 10^{10}$ cm leads to $\tau \sim \frac{10^3}{H}$. Hence roughly $1 < H \leqq 50$ gauss an be estimated which appears quite plausible for heights of 0.3 solar radii.

The emission mechanism itself is not yet satisfactorily understood. Denisse (1960) proposed a Čerenkov process for plasma waves (cf. 3.6.2) which partly are transformed into electromagnetic waves by a coupling mechanism. This would explain certain characteristics of the continuum storms, e.g. the polarization and the emission near the plasma frequency. But the explanation of the long duration and the development into type I noise storms remain as unsolved problems.

Another attempt undertaken, by Ginsburg and Zheleznyakov (1958) is based on the assumption of coherent plasma waves generated by a beam of non-thermal particles. This is essentially the same mechanism which was tempted by Takakura (1963) for the interpretation of the type I bursts.

According to Bohm and Gross (1949) the amplitudes of the plasma oscillations in a stationary equilibrium are given by the formula

$$\frac{e\,E_0}{k} = \frac{m_e}{2}\left(\frac{\omega}{k}\right)^2 \left\{\frac{16}{3} \cdot \frac{\omega_p^2}{k^2} \cdot \frac{\tau_s}{\tau_f} f'\left(\frac{\omega}{k}\right)\right\}^2 \approx \frac{m_e}{2} v_0^2 \left\{\frac{16}{3} \cdot \frac{\tau_s}{\tau_f}\left(\frac{v_0}{v}\right)^2 \frac{N_f}{N_o}\right\}^2 \tag{4.4.4.-2}$$

where k — wave number, τ_f and τ_s — mean collision times for fast and slow particles, respectively, $f'(\omega/k)$ — slope of the velocity distribution function $[f(v)]_{v=v_k}$, v_k — phase velocity of the plasma waves.

The transformation into electromagnetic waves is possible by a scattering at thermal fluctuations of the electron density. The emitted energy is given by

$$W = \frac{n_f\, e^4\, N_{sc}\, V_{sc}\, E_0^2}{6\, m_e^2\, c^3} \tag{4.4.4.-3}$$

(Ginsburg and Zheleznyakov, 1958), where N_{sc} and V_{sc} denote the density and volume of the scattering region, respectively. For the effective temperature the expression

$$T_{\text{eff}} = \frac{c^2\, W}{2\, k'\, \nu^2\, S\,(\Delta\nu)} = 8.4 \cdot 10^{-12}\, \omega_p^2\, n_f \frac{N_{sc}\, L\, v_0^{10}}{\nu^2\, \Delta\nu\, \bar{v}^8}\left(\frac{\tau_s}{\tau_f}\right)^4 \left(\frac{N_f}{N_0}\right)^4 \tag{4.4.4.-4}$$

is obtained, where k' — Boltzmann's constant, $S(\Delta\nu)$ — cross section of the scattering region for a band width $\Delta\nu$.

Using this mechanism, from typical observations ($T_{\text{eff}} = 10^8 \cdots 10^{10}$ °K) a kinetic energy density of the emitted electrons

$$P_{\text{kin}} = \frac{m_e\, v_0^2}{2} N_s \sim 2 \cdot 10^9 \quad [\text{eVcm}^{-3}] \tag{4.4.4.-5}$$

follows. For $N_f \sim 10^5$ cm^{-3} an average energy of the order of 10^4 eV per radiating electron is to be expected.

4.5. Proton flares

4.5.1. Proton flares and type IV bursts

Proton flares as the most active phenomenon of solar activity have found a remarkable practical and theoretical interest. Because without exception all proton flares are closely related to the emission of radio bursts a study of the burst phenomena provides an essential contribution to the exploration of the whole phenomenon. At first a number of characteristic conditions in solar active regions leading to proton flares with a high probability must be taken into mind:

1. Complex structure of the magnetic fields, δ- and A-configurations (no simple dependency on the magnitude of the related spot group).
2. Fast development of the related spot group, rapid increase of the longitudinal gradients of the magnetic field ($\sim$1 gauss/km) and of the whole magnetic energy content.
3. Enhancement of the electron density in coronal condensations in connection with a steep increase of the S-component of the radio emission (especially at 3 cm wavelength) and of the slowly varying soft X-ray emission (1—8 Å).
4. Enhanced burst activity in the upper cm-region.

The most characteristic features of the proton flares in all spectral regions are as follows:

1. Occurrence of a broadening double ribbon structure of the $H\alpha$-flare parallel to a neutral line of the longitudinal magnetic field often covering the penumbra of a strong sunspot. The development of this structure is initiated in the so-called Y-phase (Křivsky, 1965) which is closely related to the explosive phase of the flare-burst phenomenon. Unfortunately there is still a lack of analogeous observations of structures for the sources of the radio and X-ray bursts.
2. There is an evidence for a greater importance of the transition region from chromosphere to corona for the triggering of proton flares. Correspondingly the type IVμ-burst seems to be a good indicator for the production of high energy particles and the action of the initial process of the particle acceleration.
3. In order to get particle effects near the earth a release of the particles into the interplanetary space and a favourable position of the active region on the Sun (preferably on the western solar hemisphere) apart from the conditions of the interplanetary medium are necessary conditions. The release of the particles is accompanied with special type IV burst characteristics in the cm- and m-regions: In many cases a simultaneous strong onset of the explosive phase in both the cm- and m-region indicates a potential particle event. It should be stressed out that this phase of the type IVm burst is not necessarily connected or even identical with a moving component of the type IV event.

Though specific characteristics of the proton flares are reflected by some features of the optical and radio emission in the cm-range, the location and time interval of the main acceleration process and the release of particles must not necessarily be restricted to phenomena taking place in the deeper levels of the solar atmosphere (chromosphere, low corona). In particular the contribution of the secondary acceleration mechanisms (cf. 3.7.3—3.7.5) demands a critital analysis in order to test the possibility of a production and ejection of energetic particles in higher levels of the solar atmosphere.

From a survey of proton flares associated with type IV bursts during the last solar cycle it became clear, that besides the emission in the upper cm-range also the type IV burst components at meter wavelengths must play a specific role. Therefore it can be assumed that an effective acceleration takes place also in those levels which are responsible for the m-emission. In comparison to this, the emissions of the dm- and Dm-components appear to have not such importance as indicator for an acceleration to extremely high energies. Nevertheless these spectral ranges could serve as indicators for a release of particles from the Sun and for a storage of particles of lower energies.

Considering the time characteristics of the burst emission it is evident, that short-lived impulsive phenomena (which, properly speaking, do not belong to the type IV bursts) are more likely accompanied with an acceleration of electrons, than of heavier particles as protons. A classical example is the phenomenon of type III bursts. A sufficient strong acceleration of protons would require a cyclic action of a (perhaps secondary) acceleration mechanism over a longer time interval. For this reason the main acceleration of particles is assumed to take place in a more or less extended time interval of the main phase of the burst development displayed as the explosive phase. Also from comparisons with the arrival times of the particles at the earth it is expected, that the high energy particles are released just during this phase.

4.5.2. Results of the Proton Flare Project

The first attempt to collect all available data and characteristics of a proton event on a broad international basis was undertaken during the first Proton Flare Project period in 1966 (Švestka and Simon, 1969). Some of the most important results have been used already in the previous section. But it became clear that for a detailed study of the phenomenology and physics of proton flares more investigations are needed. Because of the complexity of the phenomenological details and physical problems only systematic studies on a greater number of such events (including weak and simple cases) seem promising to get new results. In addition more theoretical studies are needed.

4.6. Models of the development of compound bursts

4.6.1. Analysis of the typical phases of a compound burst event

Large flare events accompanied by a type IV burst emission and an expellation of particles exhibit a multitude of various types and components of the flare-burst emission involving radiations in other spectral regions which reveals the complicated picture of the whole complex event. Now we try to obtain a conception for an uniform model of the development of a representative "compound" burst containing the totality of all observed burst emissions. This task comprises a translation of the time sequence of the various burst phenomena into an unique picture describing the action of special physical processes.

A proposal for such a picture is given in table 12, where in a generalized form the development of a compound burst event is represented by five phases.

The phase of the pre-burst evolution in the widest sense starts with the formation of an active region on the Sun and the hence resulting process of the storage of energy (cf. 4.1). In a more special sense, however, the pre-burst phase consists of the development to a configuration leading to an instability immediately connected with a burst event. In special cases the pre-burst phase is connected with a heating (compression ?) of a special plasma volumen which can be indicated by precursors and gradual bursts in the radio and soft X-ray regions. If a critical limit is exceeded depending on the configuration of the magnetic field

$$E > E_{\text{crit}}(H, r, t) \,, \tag{4.6.1.-1}$$

an instability occurs which by some trigger mechanism leads to the onset of the primary acceleration process taking place during the impulsive phase of the burst event. It should be noted that the bad understanding of the nature of that instability and the triggered acceleration process is one of the most critical points for the present knowledge of the whole flare-burst complex.

The impulsive phase is often identical with the flash phase of the $H\alpha$-flare and can be regarded as the cause for the ejection of streams of fast electrons indicated by the occurrence of type III bursts. Though the question of the acceleration mechanism is not yet solved, for this mechanism a higher power law of the energy dependency is demanded than e.g. that of the Fermi-Parker mechanism which works too slowly (cf. 4.2.1).

The particles accelerated in the impulsive phase can be thought to deliver the injection spectrum for the onset of secondary acceleration mechanisms provided that this injection is strong enough and favourable conditions for an action of a secondary acceleration are present.

The nature and duration of the action of the acceleration and loss processes determine the magnitude of the energy attained by the particles within a certain spatial volume

$$E_{\text{tot}} = \left\{ \int_0^\infty N(E, t)\, E\, dE \right\}_{t \leq t_d} \tag{4.6.1.-2}$$

or in the average

$$\overline{E} = \left\{ \frac{\int_0^\infty N(E, t)\, E\, dE}{\int N(E, t)\, dE} \right\}_{t \leq t_d} \tag{4.6.1.-3}$$

In general at best during the time interval of this acceleration the possibility of a stronger ejection of energetic particles can be expected (cf. 3.7.3). Therefore

Tab. 12. Main phases of the development of compound bursts

Action \ Phase / Physical process		(0) Pre-burst phase	(1) Impulsive (flash)phase	(2) Type IV_1 burst phase	(3) Type IV_2 burst phase	(4) Recovery phase
		storage of energy, pre-heating, compression, instability	trigger action, primary acceleration, injection of particles	secondary acceleration, ejection of particles (streams)	trapping of particles, plasma cloud	expansion, decay
Particle-wave-interaction	resonant (cyclotron emission)		(non-thermal) impulsive	type IV μ type IV $m\ A_1$	tpye IV $m\ A_2$ (moving)	type IV $m\ B$ (contimuum storm)
	non-resonant	gradual		burst (quasi-thermal)		
Wave-wave-interaction	$T\ T$		burst	type IV dm burst		
	$L\ T$		type III(V) burst	type II		type I storm
Associated phenomena		(soft)	X-rays (hard) $H\ \alpha$-flare	(sub) cosmic rays	emission of corpuscular radiation	
Time scale [min]			0	3	15	30 300

it can be concluded that the escape from the Sun for the majority of the particles takes place still during the period of this acceleration.

Following the acceleration phase, radiation and other loss processes become dominating and charged particles, in particular electrons, remain trapped in special magnetic field configurations. An ascension of a magnetically isolated plasma cloud could occur, if the radius of this cloud a

$$a \gtrsim h \qquad (4.6.1.\text{-}4)$$

becomes of the order of the scale height h of the ambient medium (KAWABATA, 1966). It is not yet clear whether the release of such a plasma cloud is identical with the occurrence of moving components of the type IV bursts, for which drift velocities of $5 \cdot 10^2$—$3 \cdot 10^3$ km/s are characteristically.

Because not all type IV burst components possessing the morphological and spectral appearance of a type IVm"A" burst seem to exhibit a moving phase, a distinction between type IVmA_1 and type IVmA_2 components has been introduced (cf. 4.4.3) which is also included in table 12. The type IVmA_1 phase corresponds in time often to the type IVμ burst and special type IVDm components and may be identical with the phase of the ejection of energetic particles ("explosive phase"). The moving type IV burst as a special dynamical phenomenon is denoted as type IVmA_2 burst.

The next phase of the development of large flare-burst events corresponds to the occurrence of stationary type IVm"B" components which have been discussed in 4.4.4. The radiation causes a redistribution of the particle orbits and pitch angles resulting in an average decrease of the angles Φ (BRICE, 1964). This phase can contain either a more or less extended recovery and decrease of the burst emission or the development into a continum storm.

The thus obtained scheme of different burst phases, of course, reveals a very crude and idealized picture of the great variety of possible phenomena and the internal connection between them. In particular only the largest events exhibit

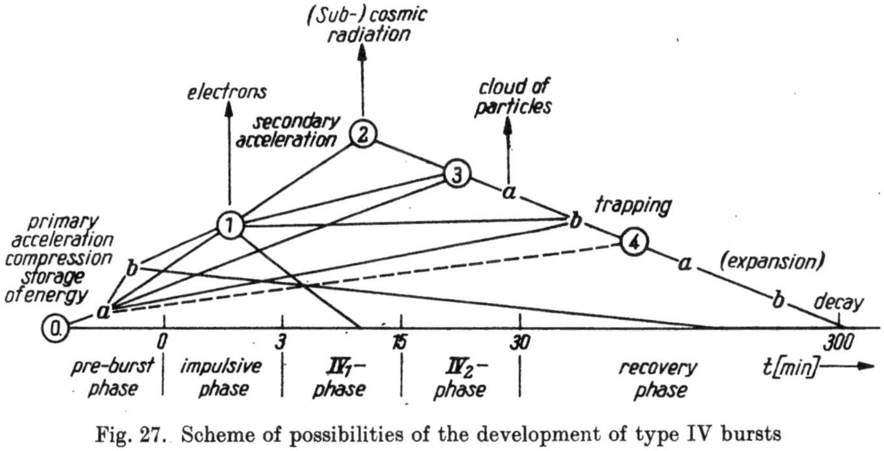

Fig. 27. Scheme of possibilities of the development of type IV bursts

a complete series of all the types and subtypes like those given in table 12. For the majority of small and moderate events only a combination of certain selected phenomena can be expected. This is, for illustration, demonstrated in fig. 27, where in completion to table 12 the different stages of the burst development are shown and the possibilities of skipping some of them in different ways.

4.6.2. On the origin of the radio emission of different spectral regions

A further important point of the interpretation of solar radio bursts comprises the complex of the question after the height level of the centres of the burst emission in the solar atmosphere, i.e. the relation mainly between the emitted spectral characteristics, solar plasma parameters, and the position of the source region. Because of the restricted possibilities of direct observations which up to now are available only for special cases as limb observations and some interferometric measurements, informations about the emitting volumen must be derived indirectly. But here difficulties arise with respect to the fact that the spectral distributions of the emitted burst radiation depend on several parameters as electron density, magnetic field strength and particle energy. The composition and variation of the actual values of the plasma parameters lead to the multitude of the observed concrete burst radiations depending on the spatial configurations and the special emission mechanism. Because of this complexity at present only rough model assumptions postulating the action of some basic physical processes are possible, which must be completed by additional informations from other phenomena in various spectral regions. In the following some general remarks on this topic should be given.

We start from a consideration of the main groups of emission mechanisms. If bremsstrahlung is presumed, which according to 3.2.5 may be responsible only for small microwave bursts, it can be supposed that because of the temperature and density dependencies of the absorption coefficient the radiation comes from similar heights as it is observed for the sources of the S-component of solar radio emission. Thus, the emission is expected to originate in the upper chromosphere and lower corona. For an inhomogeneous model of the source region the observing frequencies are related to special levels in a similar manner as it was supposed for the source regions of the S-component (e.g. NEWKIRK, 1961; KAKINUMA and SWARUP, 1962).

If gyro-synchrotron radiation is considered, for each volume element the local gyrofrequency (3.4.1.-6) determines a lower limit of the frequency interval of the emitted spectrum. This demonstrated in fig. 28 where curves of equal gyrofrequencies are given in dependence on the electron energy and the magnetic field strength for different combinations of the angles Φ and θ. It should be noted that for mean values of Φ and θ within a wide interval of non-relativistic energies the gyrofrequency varies only weakly with the energy. In a certain region a unique relation between the ν_g- and β-values is not given because one

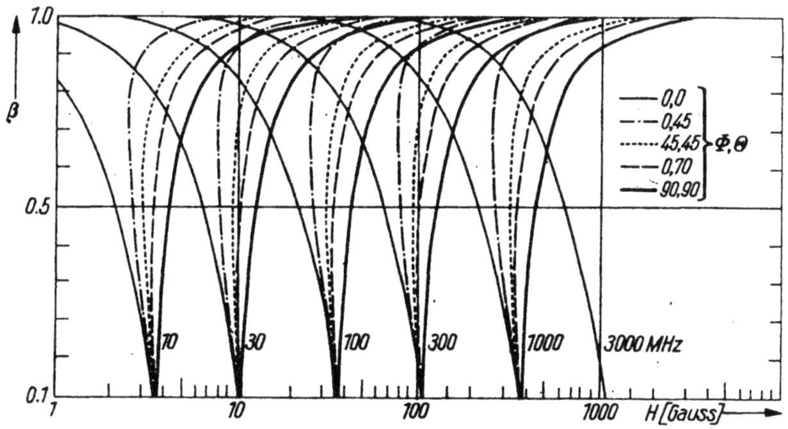

Fig. 28. Curves of equal gyrofrequency in dependence of particle energy and magnetic field for different values of the angles Φ and θ $(n_j = 1)$

value of the gyrofrequency can correspond to two different β-values. However, this ambiguity is of interest only for higher electron energies.

The connection between the spectral range and the emitting level in the solar atmosphere for gyro-synchrotron radiation and given particle energies in a first approximation is determined by the magnetic field and the electron density, both decreasing with heigth above an active region.

If the refractive index departs significantly from unity , the emission equation (3.4.1.-6) must be solved simultaneously with the corresponding dispersion relation (e.g. (3.3.1.-4)) yielding some modifications for the gyrofrequency in comparision to the case $n = 1$ especially at lower frequencies. A consequence is, that the fundamental frequency (and partly also the first harmonics) of the gyrofrequency are emitted only in a restricted range of the angle θ. For a judgement of the emission characteristics the ratio between the local plasma and gyrofrequency is of great interest. The height dependencies of the plasma frequency over an active region corresponding to the tenfold of the BAUMBACH-ALLEN distribution (cf. NEWKIRK, 1961; KUNDU, 1965) and of the gyrofrequency according to a hypothetical averaged magnetic field model belonging to a strong spot group (3000 gauss in the maximum) are shown in fig. 29. (cf. also NEWKIRK, 1967). In principle, in the solar atmosphere both cases $\nu_H > \nu_p$ and $\nu_H < \nu_p$ are possible depending on the height above the photosphere. In the example of fig. 29 three regions can be distinguished for which the ratio ν_p/ν_H is >1 or <1.

According to the proposed model of ν_p and ν_H the frequencies

$$\nu_0 = \frac{\nu_H}{2} + \left(\nu_p^2 + \frac{\nu_H^2}{4}\right)^{1/2} \qquad (4.6.2.\text{-}1)$$

denoting the zeros of the refractive index of the extraordinary wave (cf. table 7) are included in fig. 29. Furthermore the regions where $Y \ll 1$ and $X \ll 1$ are

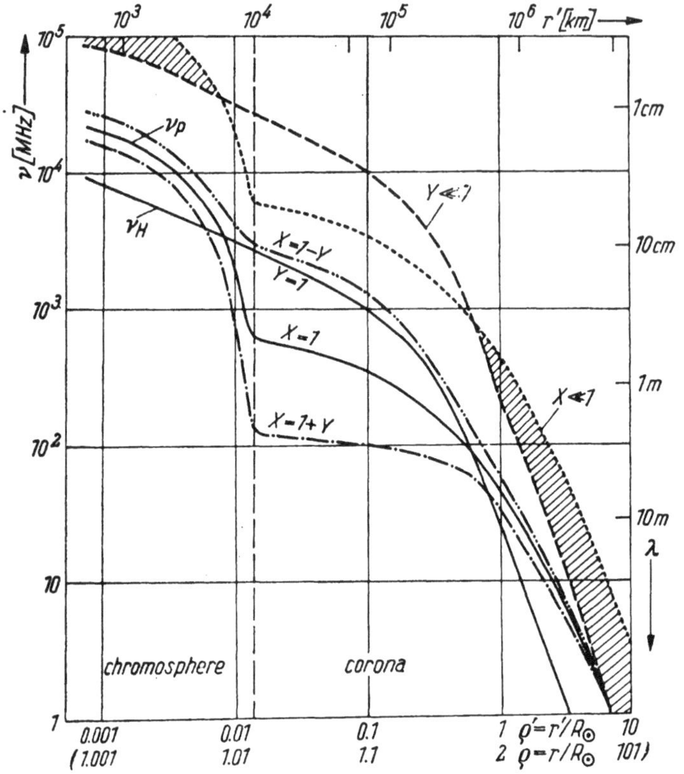

Fig. 29. Model of the height distribution of the plasma frequency and gyrofrequency in an strong active region. Shaded areas refer to potential regions of the RAZIN effect

tentatively marked by the shaded areas denoting the range of the validity of RAZIN's approximation. Taking into account the influence of the critical levels of the wave propagation together with assumptions of the mean particle energies and pitch angles, a certain judgement about the range of the probable emission levels becomes possible. However, the extension of the real emitting plasma volume and inhomogeneous distributions of the plasma parameters within it causes a superposition of the emissions from all parts of the source region and hence a broadening of the resulting burst spectrum.

Because fig. 29 represents a case of extremely high magnetic field strengths, for weaker active regions the frequency range where $\nu_H > \nu_p$ is expected to be more restricted.

4.6.3. On the estimation of plasma parameters in characteristic burst emitting regions

The main plasma parameters taking an effect on the radio burst emission are the particle kinetic energy (or equivalent temperature), the particle density, and the magnetic field strength. If it is further taken into account, that for an estimation of the intensity of the (gyro-synchrotron) radiation under simplest (i.e. homogeneous and isothermal) model assumptions for a given time the solid angle Ω of the source region and representative values of the angles Φ and θ must be known, then for a knowledge of the mean conditions in the burst source at least six independent informations are needed. A basic difficulty results from the fact that in general different observing frequencies are not related to exactly the same emitting volumes (cf. 3.3.5), so that spectral observations automatically involve the need of constructing inhomogeneous burst models and the number of necessary independent informations is increased.

Consequently as much as possible other informations must be collected including interferometric and polarimetric observations etc. But in practice this still remains insufficient. Then working hypotheses must be introduced in order to reduce the variety of possible solutions. In this case the accuracy of the obtained results depend on the nature and reliability of the proposed assumptions. In this way solar burst physics is characterized by — besides the improvement of the theoretical foundations — a successive improvement and completion of reliable informations resulting in a reduction of the fraction of not verified hypothetical assumptions. We must confess, that the present state of burst physics still contains an appreciable fraction of such not well proved model assumptions.

Concerning the knowledge of the special plasma parameters the following brief remarks can be given. The estimation of the magnetic fields was already considered in section 4.1.2. Considering the methods of an estimation of the electron density it must be noted, that the number of particles emitting gyro-synchrotron radiation is only a fraction of the whole number of particles present in the plasma volume. Usually the part of high energy particles is by some orders of magnitude lower than the density of the ambient medium. For an estimation of the whole density the influence of the medium can be taken into account, if this takes any remarkable effect. This touches the question of the energy balance which will be considered in the next section.

4.6.4. The energy exchange for different phases and spectral ranges of the flare-burst phenomenon

Considering the energy exchange of the flare-burst phenomenon some basic definitions and quantities must be taken into mind. In particular three main quantities, the mean kinetic energy per radiating particle, the kinetic energy density per volume unit, and the fraction of the energy transformed into radiation are important characteristics describing the state of the source region. The relation between these and other quantities is shown in a schematized form in table 13. The average values of these quantities can be quite different for different components of the flare-burst event. This is demonstrated in table 14 where a drafe compilation of the to-day's knowledge concerning numerical values of the average energy exchange in different components and phases of the flare-burst complex is given.

Tab. 13. Schematic representation of the basic quantities describing the energy of the solar radiation

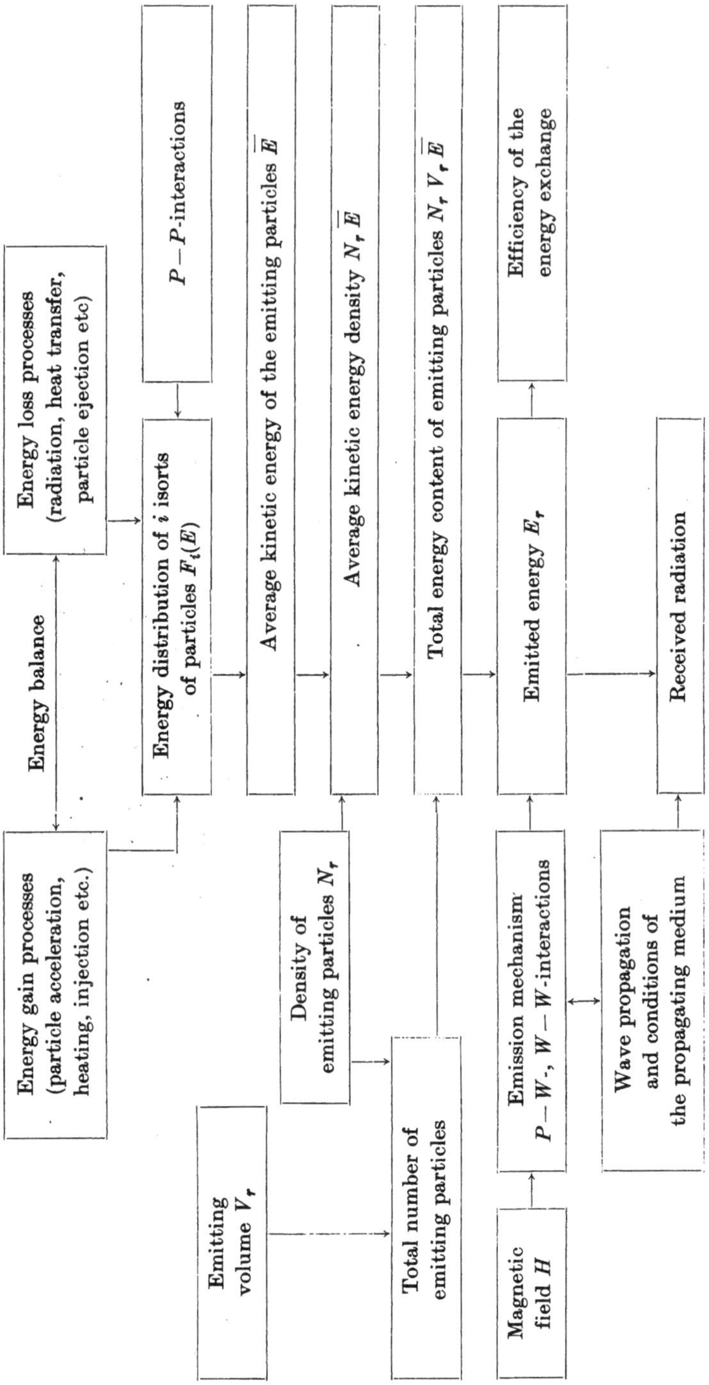

Tab. 14. Average energy content of different bursts components

	$H\alpha$-flare	X-ray-burst		cm-burst		Type IV-burst			
		soft	nord	gradual	impulsive	cm	dm	mA	mB
Average kinetic energy E of radiating particles [keV]	10^{-3}–10^{-2}	1	10–10^{3}	1	10–10^{3}	10^{2}–10^{3}	10^{2}–10^{3}	10^{2}–10^{4}	(10^{2}–10^{3})
[erg]	10^{-11}–10^{12}	10^{-9}	10^{-8}–10^{-6}	10^{-9}	10^{-8}–10^{-6}	10^{-7}–10^{-6}	10^{-7}–10^{-6}	10^{-7}–10^{-5}	(10^{-7}–10^{-6})
Energy distribution $F(E)$	therm.	qu-therm.	E^{γ}	qu-therm.	E^{γ}	quasi-monoenergetic (?)			
Total number of radiating particles $N_e V$	10^{40}–10^{42}	10^{38}–10^{39}	10^{34}–10^{36}	10^{38}–10^{39}	10^{30}–10^{32}	10^{31}–10^{33}	10^{32}–10^{33}	10^{31}–10^{36}	(10^{31}–10^{33})
T_{eff} [°K]	10^{4}–10^{5}	10^{7}	10^{8}–10^{11}	10^{7}	10^{7}–10^{11}	10^{10}	10^{10}	10^{8}–10^{13}	10^{11}
V [cm^{3}]	10^{27}–10^{29}	10^{28}–10^{29}	10^{28}–10^{29}	10^{29}–10^{30}	10^{28}–10^{30}	10^{29}–10^{30}	10^{30}	10^{30}–10^{34}	10^{30}–10^{31}
N_e [cm^{-3}]	10^{13}	10^{10}	10^{5}–10^{7}	10^{9}	10^{2}–10^{3}	10^{2}–10^{3}	10^{2}–10^{3}	10–10^{2}	(10–10^{2})
H_{max} [Gauß]	$3 \cdot 10^{3}$	10^{3}	10^{3}	$5 \cdot 10^{2}$	$5 \cdot 10^{2}$	$5 \cdot 10^{2}$	50–100	5–10	1–10
Kinetic energy density $E N_e$ [erg cm^{-3}]	10–100	10	10^{-1}–10^{-2}	~ 1	10^{-4}–10^{-5}	10^{-3}–10^{-4}	10^{-3}–10^{-4}	10^{-4}–10^{-5}	(10^{-4}–10^{-5})
Total energy content $E N_e V$ [erg]	10^{28}–10^{31}	10^{29}–10^{30}	10^{27}–10^{30}	10^{29}–10^{30}	10^{24}–10^{25}	10^{26}–10^{27}	10^{27}–10^{28}	10^{26}–10^{29}	(10^{26})
Energy transformed into radiation [erg]	10^{28}–10^{31}	10^{26}–10^{29}	10^{25}–10^{29}	10^{23}–10^{24}	10^{22}–10^{23}	$\sim 10^{25}$	$\sim 10^{23}$	$\sim 10^{23}$	$\sim 10^{23}$

The lowest kinetic energy of about 10^{-3} keV is represented by those particles which emit the optical $H\alpha$-flare radiation. In spite of this low energy, due to the higher density the energy density is higher than in the source regions of the X- or radio burst emission. Besides this there is a high efficiency for the transformation of the kinetic energy into $H\alpha$-radiation so that the main part of the energy emitted into electromagnetic radiation by solar burst events comes from the optical (and possibly UV-) flare emission. In comparison to this, the amount of the sporadic X-ray emission is by several orders of magnitude lower. But the energy of the emitting particles for the X-ray emission is some orders of magnitude greater than that of those of the $H\alpha$-radiation.

Similar to the X-ray bursts also for the microwave bursts particle energies in a wide interval between 1 and 10^3 keV are characteristically. Correspondingly also the numbers of emitting particles and the kinetic energy density are variable within a wide range. A distinction between two large characteristic energy ranges can be made regarding the quasi-thermal (gradual) bursts and the non-thermal impulsive of even type IV-like events.

The type IVmA bursts probably can be considered as the strongest burst component with respect to the energy content of fast particles as well as regarding the relatively large emitting volume. But due to the relatively small spectral range of the related emission the energy transformed into radiation for this component has no exceptional characteristics. For the type IVmB burst it is yet difficult to get conclusive data because the amount of plasma radiation and the efficiency of the coupling mechanisms is not well understood, the most probable values are given in parentheses in table 14.

From a comparison between the kinetic energy densities with the corresponding magnetic energy densities

$$\frac{H^2}{8\pi} = 5.1 \cdot 10^{-4}\, \nu_{H[\mathrm{MHz}]} \qquad [\mathrm{erg\ cm^{-3}}] \qquad (4.6.4.\text{-}1)$$

it is expected that in the burst emitting regions the average kinetic energy densities are typically smaller than the magnetic ones. But an escape of particles is possible for particles having gyroradii greater than a characteristic dimension of the trapping region. It may be concluded, that concerning the output of the energy from any flare-burst event two principal kinds of source regions must be distinguished: One region, from where the energy comes out mainly in form of optical (and UV) radiation (low energy particles) and another region with a relatively small output of electromagnetic radiation (radio and X-ray ranges), but where the energy is released predominantly in form of particle emission.

It was already mentioned, that the energy of maximal 10^{33} erg exchanged in a big flare-burst event is only a small fraction of the whole magnetic energy content of an active region. Similarily the energy emitted by the burst radiation is a negligible fraction of the total energy exchanged in a flare-burst event. Nevertheless, the burst radiation is an important indicator for the highest energies of the particles inside the flare-burst volume and contains informations about the extension of the flare disturbance in greater heights of the solar atmosphere (corona), which otherwise hardly could be obtained.

4.7. Problems of the forecasts of solar flares and bursts

The aim of solar burst physics is not accomplished in a mere interpretation of observed events, only the look into future solar activity provides a test for the significance of the explanation of the observed phenomena, the reliability of peculiar detected laws, and the correctness of the introduced model assumptions. For this reason now we consider the most difficult question of burst physics — the forecast of solar flare-burst events. Besides the proof of a right understanding of the flare-burst complex, such a forecast-especially of major events as proton flares — is of a great practical interest in several directions (cf. Švestka, 1968; Simon, 1969; Severny, 1969; Severny and Steshenko, 1969).

The problem of flare forecast can be devided into four general parts:

a) Long-term forecasts over several solar rotations
b) Medium-term forecast for about one solar rotation
c) Short-term forecast for several days
d) Judgement of the possibility of a particle event from the characteristics of a special flare-burst event in progress still before the arrival of the particles on the earth.

Points a) and b) contain most difficult and yet unsolved questions. Without knowledge of the exact cause for the formation of a concrete solar centre of activity with its individual structure and development, up to now it is possible to give only some general ideas about principal possibilities of a long-term forecast which are based on a more or less regular behaviour (several maxima) during the solar cycle and the preference of special intervals of Carrington longitudes during a certain period for proton active centres (Dodson-Prince and Hedeman, 1968).

For the medium-term forecast a knowledge of the activity on the far side of the Sun is desired. Indirect methods indicating energetic processes on the far side are not yet systematically developed and artificial solar satellites will be available only in future.

Concerning the short-term forecast the situation is somewhat better. A number of characteristics is investigated, which can be used for a well organized international forecasting service:

a) The photospheric magnetic field configuration (gradients, polarities, structural details)
b) The occurrence of flares of special structures in complex spot groups
c) The velocity of the development of an active region

d) The position on critical CARRINGTON longitudes.

Because the main part of the acceleration to relativistic particle energies takes place in the source region of the radio emission, we consider additional criteria basing on radio observations:

e) Development and spectrum of the S-component
f) Rates of the occurrences of bursts in the lower cm-region and of X-ray bursts
g) Repeating of special type IV burst events.

The spectrum of the S-component of the solar radio emission is influenced by the storage of kinetic and magnetic energy in active regions (cf. 4.1.3) and indicates the realization of preconditions necessary for the occurrence of flares and bursts. There is a close correlation between the emission of the S-component and the frequency of occurrence of bursts especially in the microwave region. But it would be dangerous to consider these connections as too simple, because neither the application of the method of summation curves of daily means of the S-component nor the existence of a spectrum increasing towards higher frequencies in the cm-region leads to definite results. The ratio between the fluxes of the S-component at 3 and 20cm wavelength (or familiar wavelengths) provides a rough estimation of the possibility of large outbursts in special time intervals. For this purpose a high interdiurnal and long-term accuracy of the patrol measurements is needed (absolute receiver calibrations!). Statistical fluctuations of the measured values and systematic errors injure the conclusions.

The number of impulsive bursts per day (impulsive cm- and X-ray bursts, type III bursts) gives some indication about the tendency of an active region to produce instabilities. From the spectrum (bandwidth, maximum frequency, etc). and the morphology (complexity, impulsiveness, etc.) especially of the microwave bursts the effectiveness of the acceleration processes can be deduced, which is of interest for a judgement of the potential development of an active region.

For illustration in fig. 30 a distribution of the ratio of microwave burst to flare activity is shown in dependence of the type of the related spot groups. This ratio increases with growing complexity of the spot development stressing out, that additionally to the optical observations the radio emission contains further informations about energetic processes in active regions (KRÜGER, 1969). Analogeously also repeating tendencies of type IV bursts contain informations about the possibility of a production of large events in certain active regions, but naturally in general the number of such events is rather small so that individual events may considerably depart from an "average" behaviour.

In summary the conclusion can be drawn, that besides the optical observations of photospheric magnetic fields, the radio emission gives complementary indications of potential tendencies of a production of energetic events, which at least for the short-term forecast should be taken into consideration.

In predicting proton emissions coming near to the earth (PCAs) from the characteristics of a flare-burst event still in progress, the type IV burst has been

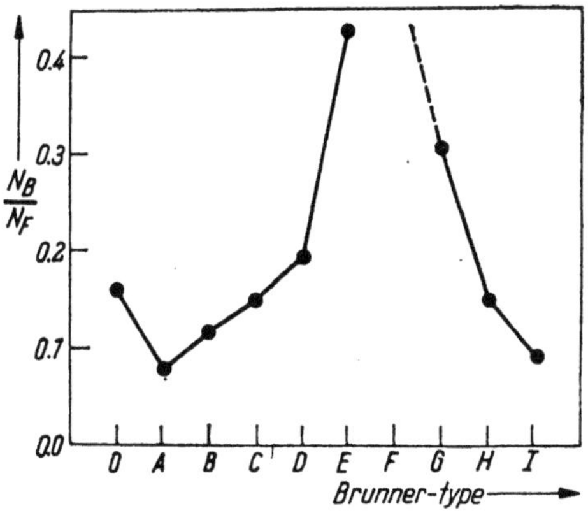

Fig. 30. Ratio of daily numbers of cm-bursts N_B to those of flares N_F in dependence of the type of the related spot group

proved as valuable indicator. The following criteria can be used for a prediction of major PCAs:

a) Existence of type IVμ and IVmA_1 components
b) Sufficiently great maximum flux in the 10 GHz range (exceeding 10^3 s.u.) and large ratios of $S_{10000,\ max}/S_{3000,\ max}$ (>1)
c) Sufficiently long duration (in contrast to merely impulsive bursts)
d) Sufficiently strong energy spectrum (exceeding 10^{-17} Joules m^{-2} Hz^{-1}) and total energy (exceeding 10^{-6} Joules m^{-2}) in the 10–1 GHz range
e) preference of western heliographic longitudes (cf. Böhme, 1969; Straka and Barron, 1969).

5. Summary and conclusions

At the end of this book we are tempted to reconsider the achieved results and to think over the next steps necessary in the research of burst physics for future work.

As stated in the first chapter, in the field of observations basic properties of the burst radiation have been investigated for many years by means of statistical analyses and extensive studies of single events. In future (a) a standardization of the measurements on an international basis in order to get homogeneous series of comparative data over long periods, (b) an extension and improvement of special measurements (in particular enlargement of the spatial and temporal resolution powers), and (c) complex problem oriented experiments directed on the solution of special questions possesing more or less a key position in solar burst physics certainly will obtain greater importance. Such key questions could be as follows

- detailed explanation of the main processes of storage of energy in active regions preceding flare burst events
- judgement of different kinds of instabilities in active regions
- investigation of distinct forms of trigger actions
- clarification of the nature of the acceleration processes in connection with burst events especially for the initial phase
- knowledge of the structure and evolutional tendencies in all levels of an active region with regard to a potential generation of large (type IV-like) burst events.

In contrast to these more or less open questions the theory of the emission processes — as treated in chapter 3 — appears better developed. Different steps of an approximation could be distinguished:

- The vacuum theory (single particle approximation)
- the magneto-ionic theory (cold plasma)
- the kinetic theory (warm and hot plasma).

Nevertheless, the elaboration of the theoretical fundamentals is not finished in all respects even in such consolidated fields like the theory of gyro-synchrotron radiation as it is demonstrated by the influence of the medium. Even the most advanced approach to the theory of emission mechanisms, the kinetic theory, demonstrates serious limitations of our knowledge e. g. with respect to a missing of a theory of strong absorption processes and a comprehensive picture of radiation instabilities. Also the actual influence of different kinds of plasma waves is not fully known. Some tasks for the future are the following:

- improvement of the theory of emission mechanisms especially with regard to a broader application of the kinetic theory (non-equilibrium), and the consideration of non-linear effects (instabilities)
- elaboration of a detailed surrey of possible generation processes of plasma waves and their coupling mechanisms with longitudinal and transveral wave modes in a turbulent warm and hot plasma
- quantitative tests of the theoretical fundamentals and solar conditions by catalogues of model computations of radiation characteristics including polarization effects based on different distributions of solar plasma parameters
- calculation of dynamical models for the burst morphology applying results of an explanation of the basic question of particle energy gain and loss processes.

Because for theoretical burst models — especially under consideration of inhomogeniety and unisotropy of the source region — the knowledge of a certain number of physical parameters is needed, the successive improvement of interpretation and theory requires a growing amount of useful informations. Therefore, besides the theoretical tools, also the development of new techniques of observation is of crucial importance for a support or refusing of the results of the present attempts of painting a physical picture of solar burst phenomena.

Bibliography

Acton, L. W. (1968): Astrophys. J. **152**, 305

Akabane, K. (1958): Ann. Tokyo Astron. Obs. 2nd. ser. IV, 57

Akinyan, S. T. and Mogilevsky, E. I. (1961): Geomagn. i Aeronom. **1**, 156

Alfvén, H. (1941): Ark. mat. astr. och fys. **27A**, No. 25

Alfvén, H. (1965): Rev. Mod. Phys. **37**, 652

Alfvén, H. and Carlquist, P. (1967): Solar Physics **1**, 220

Alfvén, H. and Fälthammer, C. C. (1963): Cosmical Electrodynamics, Clarendon Press, Oxford

Allen, C. W. (1963): Astrophysical Quantities, 2nd. ed. Athlone Press, London

Allen, C. W. (1965): Space Sci. Rev. **4**, 91

Allis, W. P., Buchsbaum, J. S. and Bers, A. (1963): Waves in Anisotropic Plasmas, M.I.T. Press, Cambridge, Mass.

Anderson, K. A. and Winckler, J. R. (1964): J. Geophys. Res. **67**, 4103

Antalova, A. (1965): Bull. Astron. Inst. Czechoslovakia, **16**, 32

Antalova, A. (1967): Bull. Astron. Inst. Czechoslovakia, **18**, 61

Arnoldy, R. L., Kane, S. R., and Winckler, J. R. (1967): Solar Physics **2**, 171

Arnoldy, R. L., Kane, S. R., and Winckler, J. R. (1968): Astrophys. J. **151**, 711

Athay, R. G. and Moreton, G. E. (1961): Astrophys. J. **133**, 935

Avignon, Y., Martes, M.-J., and Pick, M. (1966): Ann. d'Astrophys. **29**, 33

Bailey, D. K. (1964): Planet. Space Sci. **12**, 495

Banin, W. G. (1965): Izv. Krymsk. Astrofiz. Obs. **34**, 252

Banin, W. G. (1966): Izv. Krymsk. Astrofiz. Obs. **35**, 190

Basler, R. P. and Owren, L. (1964): Scient. Rep. Geophys. Inst. Univ. UAG — R 152

Bekefi, G. (1966): Radiation Processes in Plasma, John Wiley, New York

Bekefi, G., Hirshfield, J. L., and Brown, S. C. (1961): Phys. Rev. **122**, 1037

Bernstein, I. B. (1958): Phys. Rev. **109**, 10

Blizard, J. B. (1963): Bull. Amer. Phys. Soc. **8**, 113

Böhme, A. (1962): Gerlands Beitr. Geophys. **71**, 211

Böhme, A. (1964): Beitr. Plasmaphys. **4**, 137

Böhme, A. (1969): Gerlands Beitr. Geophys. **3**, 193 and **4**, 283

Böhme, A. and Krüger, A. (1967): Suppl. Ser. Observ. Data HHI, Vol. I, No. 1 and No. 2

Böhme, A. and Krüger, A. (1968): Suppl. Ser. Solar Data HHI, Vol. I, No. 3

Bohm, D. and Gross, E. P. (1949): Phys. Rev. **75**, 1851 and 1864

Boischot, A. (1957): Comptes Rendus Ac. Sci. Paris **244**, 1326

Boischot, A. (1958): Ann. d'Astrophys. **21**, 273

Boischot, A. and Clavelier, B. (1967): Astrophys. Letters **1**, 7

Boischot, A. and Clavelier, B. (1968): Ann. d'Astrophys. **31**, 445

Boischot, A. and Denisse, J. F. (1957): Comptes Rendus Ac. Sci. Paris **245**, 2194

Boischot, A. and Pick, M. (1962): Phys. Soc. Japan **17**, Suppl. A-II, 203

Boulin, J. D. and Simon, M. (1969): Solar Physics **9**, 183

Booker, M. G. (1936): Proc. Roy. Soc. London A, **155**, 235
Borovik, V. N., Gelfreikh, T. B., Peterova, N. G., and Krüger, A. (1968): Solnechnye Dannye No. 9, 101
Brandstatter, J. J. (1963): An introduction to waves, rays and radiation in plasma media, McGraw-Hill, New York
Brice, N. (1964): J. Geophys. Res. **69**, 4515
Bray, R. J. and Loughhead, R. E. (1964): Sunspots, Chapman and Hall Ltd., London
Bruce, C. E. R. (1966): Observatory **86**, 82
Bruzek, A. (1964): J. Geophys. Res. **69**, 2386
Bryant, D. A., Cline, T. L., Desai, U. D., and McDonald, F. B. (1965): Phys. Rev. Letters **14**, 481
Budden, K. G. (1961): Radio waves in the ionosphere, Univ. Press. Cambridge
Bumba, V. and Howard, R. (1965): Astrophys. J. **142**, 796
Burkhardt, G., Fahl, L. and Larenz, R. W. (1961): Z. Phys. **161**, 380
Byrne, F. N., Ellison, M. A., and Reid, J. H. (1964): Space Sci. Rev. **3**, 319

Carlquist, P. (1969): Solar Physics **7**, 377
Carmichael, H. (1962): Space Sci. Rev. **1**, 28
Castelli, J. P., Aarons, J., and Michael, G. A. (1967): J. Geophys. Res. **72**, 5491
Chandrasekhar, S. (1950): Radiative Transfer, Univ. Press. Oxford
Chang, D. B. (1962): Phys. Fluids **5**, 1564
Chapman, S. and Cowling, T. G. (1960): The Mathematical Theory of Non-uniform Gases, Univ. Press. Cambridge 2nd. ed.
Charmichael, H. (1964) in: The Physics of Solar Flares, ed. by W. N. Hess, Washington
Chiuderi, C. and Drago-Chiuderi, F. (1966): Half-Life of Microwave Radio Bursts and Determination of Solar Magnetic Fields, Paper presented at the Annual Meeting of Societa Italiana di Fizica, Trieste, October 24–28, 1966
Chubb, T. A., Kreplin, R. W., and Friedeman, H. (1966): J. Geophys. Res. **71**, 3611
Cohen, M. N. (1958): Proc. Inst. Radio Engrs. **46**, 172
Cohen, M. N. (1960): Astrophys. J. **131**, 664
Cohen, M. N. (1961): Astrophys. J. **133**, 978 and Phys. Rev. **123**, 711
Cohen, M. N. and Dwarkin, M. (1961): J. Geophys. Res. **66**, 411
Coleman, P. J. (1961): J. Geophys. Res. **66**, 1351
Coppi, B. (1964): Ann. Phys. **30**, 178
Covington, A. E. (1951): J. Roy. Astr. Soc. Canada **45**, 49
Covington, A. E. (1969): J. Roy. Astr. Soc. Canada **63**, 125
Covington, A. G. and Harvey, G. A. (1961): Nature, London **192**, 152
Cowling, T. G. (1945): Proc. Roy. Soc. A, **183**, 453
Cowling, T. G. (1957): Magnetohydrodynamics, Interscience Publ., New York
Crawford, F. A. (1965): Nucl. Fusion 5, 73 and Radio Science **69D**, 789

Daene, H. (1967): Astron. Nachr. **289**, 189
Daene, H., Křivsky, L. and Krüger, A. (1966): Mber. Dt. Akad. Wiss. Berlin **11**, 807
Daene, H. and Krüger, A. (1966): Astron. Nachr. **289**, 105 and 117; Mber. Dt. Akad. Wiss. Berlin **8**, 47
Das Gupta, M. K. and Basu, D. (1965): J. Atm. Terr. Phys. **27**, 1029
Denisse, J. F. (1960): Inf. Bull. Sol. Radio Obs. No. 4, 3
Denisse, J. F. et Delcroix, J. L. (1961): Propagation des ondes dans le plasma, Dunod, Paris

Dodson, H. W. and Hedemann, E. R. (1966): Astrophys. J. **145**, 224
Dodson-Prince, H. W. and Hedemann, R. (1968) in: Structure and Development of Solar Active Regions, ed. by K. O. Kiepenheuer, Reidel Publ. Dordrecht
Dorman, L. I. and Miroshnichenko, L. I. (1968): Solar Cosmic Rays, Nauka Moscow (in russ.)
Drummond, W. E. and Rosenbluth, M. N. (1960): Phys. Fluids **3**, 45
Dungey, J. W. (1953): Phil. Mag. **44**, 725
Dungey, J. W. (1958a) in: Electromagnetic Phenomena in Cosmical Physics, ed. by B. Lehnert, Univ. Press Cambridge
Dungey, J. W. (1958b): Cosmic Electrodynamics, Univ. Press Cambridge
Dupree, T. M. (1964): Phys. Fluids **7**, 923
Dvoryashin, A. S. (1962): Izv. Krymsk. Astrofiz. Obs. **28**, 305

Eidmann, V. J. (1958): Zhurn. exp. teoret. fiz. **34**, 131
Eidmann, V. J. (1959): Zhurn. exp. teoret. fiz. **36**, 1335
Elgarøy, Ø. and Eckhoff, H. K. (1966): Astrophys. Norvegica **10**, 127
Elliott, H. (1964): Planet. Space Sci. **12**, 657
Elliott, H. (1969) in: Solar Flares and Space Research ed. by C. de Jager and Z. Švestka, North Holland Publ. Amsterdam
Ellis, G. R. A. (1964): Aust. J. Phys. **17**, 63
Ellison, M. A. (1963): Quart. J. Roy. Astr. Soc., London **4**, 62
Ellison, M. A., McKenna, S. M. P. and Reid, J. H. (1960): Dunsink Obs. Publ. **1**, No. 1 and Month. Not. Roy. Astr. Soc. **122**, 491
Ellison, M. A. and Reid, J. H. (1964) in: Res. in Geophys. Vol. 1
Elwert, G. (1948): Z. Naturforsch. **3a**, 477
Elwert, G. (1961): J. Geophys. Res. **66**, 391
Epstein, R. I. and Feldman, P. A. (1967): Astrophys. J. **150**, L109
Evans, K. and Pounds, K. A. (1968): Astrophys. J. **152**, 319

Fan, C. Y., Pick, M., Ryle, R., Simpson, J. A., and Smith, D. R. (1968): J. Geophys. Res. **73**, 1555
Fermi, E. (1949): Phys. Rev. **75**, 1169
Fermi, E. (1954): Astrophys. J. **119**, 1
Fichtel, C. E. and McDonald, F. B. (1967): Ann. Rev. Astron. Astrophys. **5**, 351
Field, G. B. (1956): Astrophys. J. **124**, 555
Field, G. B. (1960): J. Geophys. Res. **65**, 1661
Field, G. B. (1961): J. Geophys. Res. **66**, 1395
Fireman, E. L. (1962): J. Geophys. Res. **67**, 4890
Flamm, E. J. and Lingenfelter, R. E. (1964): Science **144**, 1566
Fletcher, J. D. (1964): J. Amer. Inst. Aeronautics Astronautics, **2**, 2193
Försterling, K. (1942): Höchstfrequ. u. Elektroakustik **59**, 10
Fokker, A. D. (1960): Doctoral Thesis, Leiden
Fokker, A. D. (1961): Arkiv för Geofys. **3**, 459
Fokker, A. D. (1963): Space Sci. Rev. **2**, 70
Fokker, A. D. (1966): Bull. Astr. Inst. Netherlands, Suppl. **1**, 309
Fokker, A. D. (1967): Solar Physics **2**, 316
Fritzova-Švestkova, L. and Švestka, Z. (1966): Bull Astr. Inst. Czechoslovakia **17**, 249
Fritzova-Švestkova, L. and Švestka, Z. (1967): Solar Physics **2**, 87
Friedman, H. (1963): Ann. Rev. Astron. Astrophys. **1**, 59

Friedman, H. (1969): in Solar Flares and Space Research ed. by C. de Jager and Z. Švestka, North-Holland Publ. Amsterdam, p. 87
Fürstenberg, F. (1969): Inf. Bull. Solar Radio Obs. No. 25, 6
Furth, H. P., Kileen, J. and Rosenbluth, M. N. (1963): Phys. Fluids **6**, 459
Fung, P. C. W. (1966): Planet. Space Sci. **14**, 335 and 469; Aust. J. Phys. **19**, 489
Fung, P. C. W. (1967): Plasma Physics **9**, 94
Fung, P. C. W. and Yip, W. K. (1966): Aust. J. Phys. **19**, 759
Fung, P. C. W. (1968): Canad. J. Physics **46**, 1073
Fung, P. C. W. (1969): Plasma Phys. **11**, 285

Gelfreikh, G. B. (1962a): Solnechnye Dannye 1962, No. 5, 67
Gelfreikh, G. B. (1962b): Solnechnye Dannye 1962, No. 3, 50
Gershman, B. N. (1960): Soviet. Phys. JETP **11**, 657
Ginsburg, V. L. (1951): Dokl. Akad. Nauk SSSR **76**, 377
Ginsburg, V. L. (1958) in: Progress in Elementary Particles and Cosmic Ray Physics IV, ed. by J. G. Wilson and S. A. Wouthuysen, North-Holland Publ. Amsterdam
Ginsburg, V. L., Sazonov, V. N., and Syrovatsky, S. I. (1968): Usp. fiz. nauk, **94**, 63
Ginsburg, V. L. (1960): The propagation of electromagnetic waves in plasma (in russ.), Moscow (2. ed. 1967) (1964 – engl. transl. Pergamon Press)
Ginsburg, V. L. and Ozernoy, S. M. (1966): Izv. vys. ucheb. zav. Radiofizika **9**, 221; Astrophys. J. **144**, 599
Ginsburg, V. L. and Syrovatsky, S. I. (1963): The Origin of Cosmic Rays (in russ.), Moscow
Ginsburg, V. L. and Syrovatsky, S. I. (1965): Usp. fiz. nauk **87**, 65; Ann. Rev. Astr. Astrophys. **3**, 297
Ginsburg, V. L. and Syrovatsky, S. I. (1966): Usp. fiz. nauk **88**, 485
Ginsburg, V. L. and Zheleznyakov, V.V. (1958): Astron. Zhurn. 35, 694; Izv. vys. ucheb. zav., Radiofizika **1**, 59
Ginsburg, V. L. and Zheleznyakov, V. V. (1959): Astron. Zhurn. **36**, 233
Ginsburg, V. L. and Zheleznyakov, V. V. (1961): Astron. Zhurn. **38**, 3
Giovanelli, R. G. (1946): Nature **158**, 81
Giovanelli, R. G. (1947): Month. Not. Roy. Astr. Soc. **107**, 338
Giovanelli, R. G. (1948): Month. Not. Roy. Astr. Soc. **108**, 163
Gnevyshev, M. N. (1963): Astron. Zhurn. **40**, 401
Gnevyshev, M. N. and Křivsky, L. (1966): Astron. Zhurn. **43**, 385
Gold, T. and Hoyle, F. (1960): Month. Not. Roy. Astr. Soc. **120**, 89
Gopasyuk, S. I. (1964): Izv. Krymsk. Astrofiz. Obs. **32**, 14
Gopasyuk, S. I. (1967): Izv. Krymsk. Astrofiz. Obs. **36**, 56
Gopasyuk, S. I., Erushev, N. N., and Neshpor, Y. I. (1967): Izv. Krymsk. Astriofiz, Obs. **37**, 109
Gopasyuk, S. I. and Křivsky, L. (1967): Solar Physics **1**, 145
Green, R. M. and Sweet, P. A. (1967): Astrophys. J. **147**, 1153
Gurevich, A. V. (1960): Zhurn. exp. teoret. fiz. **38**, 1597
Gurevich, A. V. (1968): Zhurn. exp. teoret. fiz. **54**, 522

Hachenberg, O. and Krüger, A. (1959): J. Atm. Terr. Phys. **17**, 20
Hachenberg, O. and Krüger, A. (1964): Z. Astrophys. **59**, 261
Hachenberg, O. and Wallis, G. (1961): Z. Astrophys. **52**, 42
Hagen, J. P. and Barney, W. H. (1968): Astrophys. J. **153**, 275
Haurwitz, M. W. (1968): Astrophys. J. **151**, 351

Heald, M. A. and Wharton, C. B. (1965): Plasma Diagnostics with Microwaves, John Wiley, New York
Heitler, J. W. (1954): The Quantum Theory of Radiation, 2nd ed. Clarendon Press, Oxford
Helms, B. (1966): Unpublished
Hertweck, F. and Schlüter, A. (1957): Z. Naturforsch. **12a**, 844
Hettner, F. (1958): Z. Phys. **150**, 182
Hirth, W. (1963): Dissertation, Humboldt-Universität, Berlin
Hirth, W. (1967): Z. Astrophys. **65**, 48
Hirth, W. and Wallis, G. (1966): Beitr. Plasmaphys. **6**, 27
Holt, S. S. and Ramanaty, R. (1969): Solar Physics 8, 119
Howard, R. (1963): Astrophys. J. **138**, 1312

Ikhsanova, V. N. (1960): Izv. Astr. Obs. Pulkovo **21**, No. 164, 62
Ivanenko, P. and Sokolov, A. (1953): Klassische Feldtheorie, Akademie-Verlag, Berlin

Jackson, J. D. (1962): Classical Electrodynamics, John Wiley, New York
Jäger, F. W. (1966) in: Atti del Convegno sui Campi Magnetici Solari, Firenze
Jäger, J. and Wallis, G. (1961): Beitr. Plasmaphys. **1**, **44**
Jaeger, J. C. and Westfold, K.C. (1950): Aust. J. Sci. Res. **A3**, 376
Jager, C. De (1959): Structure and Dynamics of the Solar Atmosphere, in: Handbuch der Physik, Herausg. S. Flügge Bd. 52, Springer-Verlag
Jager, C. De (1960) in: Space Research I, ed. by K. Kallmann Bijl, North-Holland Publ., Amsterdam
Jager, C. De (1962): Space Sci. Rev. **1**, 487
Jager, C. De (1964) in: Research in Geophysics, ed. by H. Odishaw, Vol. 1, M.I.T. Press
Jager, C. De (1965) in: Introduction to Solar-Terrestrial Relations, ed. by J. Ortner and H. Maseland, Reidel Publ. Dordrecht
Jager, C. De (1966) in: The Fine Structure of the Solar Atmosphere, ed. by K. O. Kiepenheuer, Steiner-Verlag, Wiesbaden
Jager, C. De and Švestka, Z. (ed.) (1969): Solar Flares and Space Research, North Holland Publ. Amsterdam
Jaggi, R. K. (1965) in: Stellar and Solar Magnetic Fields, IAU-Symposium No. 22, ed. by R. Lüst, North-Holland Publ., Amsterdam

Kai, K. (1963): Publ. Astr. Soc. Japan **15**, 195
Kai, K. (1965): Publ. Astr. Soc. Japan **17**, 294 and 309
Kai, K. (1967): Proc. Astron. Soc. Australia **1**, 49
Kakinuma, T. and Swarup, G. (1962): Astrophys. J. **136**, 975
Kakinuma, T. and Tanaka, H. (1961): Proc. Res. Inst. Atm. Nagoya Univ. 8, 39
Kapitsa, S. P. (1960): Zhurn. exp. teoret. fiz. **39**, 1367
Kaplan, S. A. and Tsytovich, V. N. (1968): Astrofizika Akad. Nayk Arm. SSR 4, 337 and preprint No. 65 FIAN Moscow
Kawabata, K. (1960): Rep. Ionosph. Space Res. Japan **14**, 405
Kawabata, K. (1963) in: The Solar Corona, IAU-Symp. No. 16, ed. by J. W. Evans, Acad. Press New York
Kawabata, K. (1964): Publ. Astron. Soc. Japan **16**, 30
Kawabata, K. (1966): Rep. Ionosph. Space Res. Japan **20**, 107
Kellermann, K. I. (1964): Astrophys. J. 140, 969
Kellermann, K. I. (1966): Astrophys. J. **146**, 621

Kiepenheuer, K. O. (1965) in: The Solar Spectrum, ed. by C. de Jager, Reidel Publ. Dordrecht
Kiepenheuer, K. O. (ed.) (1968): Structure and Development of Solar Active Regions (IAU-Symposium No. 35), Reidel Publ. Dordrecht
Kippenhahn, R. and Schlüter, A. (1957): Z. Astrophys. **43**, 36
Kleczek, J. and Olmr, J. (1967): Bull. Astr. Inst. Czechoslovakia **18**, 68
Kleczek, J., Olmr, J., and Krüger, A. (1968): Bull. Astr. Inst. Czechoslovakia **19**, 190 and Mber. Dt. Akad. Wiss. Berlin **10**
Kolomensky, A. A. (1953): Zhurn. exp. teoret. fiz. **24**, 167
Kolomensky, A. A. (1956): Dokl. Akad. Nauk SSSR **106**, 982
Komesaroff, M. (1958): Aust. J. Phys. **11**, 201
Konyukov, M. V. (1968): Geomagn. i Aeronom. **8**, 609
Kopecky, M. and Křivsky, L. (1966): Bull. Astr. Inst. Czechoslovakia **17**, 360
Korchak, A. A. (1963): Astron. Zhurn. **40**, 994
Korchak, A. A. (1965): Geomagn. i Aeronom. **5**, 32 and 601
Korchak, A. A. and Syrovatsky, S. I. (1961): Astron. Zhurn. **38**, 885
Korolkov, D. V., Soboleva, N. S., and Gelfreikh, G. B. (1960): Izv. Astr. Obs. Pulkovo **221**, 81
Kourganoff, V. and Busbridge, I. W. (1957): Basic Methods in Transfer Problems, Oxford
Kramers, H. A. (1923): Phil. Mag. **46**, 836
Křivsky, L. (1965): Bull. Astr. Inst. Czechoslovakia **16**, 27
Křivsky, L. (1968) in: Structure and Development of Solar Active Regions, IAU-Symp. No. 35, ed. by K. O. Kiepenheuer, Reidel Publ. Dordrecht
Křivsky, L. (1969): Bull. Astr. Inst. Czechoslovakia **20**, 139 and 163
Křivsky, L. and Krüger, A. (1966): Bull. Astr. Inst. Czechoslovakia **17**, 243
Krüger, A. (1961): Dissertation, Humboldt-Universität Berlin
Krüger, A. (1962a): Mber. Dt. Akad. Wiss. Berlin, **4**, 97
Krüger, A. (1962b): Z. Astrophys. **55**, 137
Krüger, A. (1963): Astron. Nachr. **287**, 119
Krüger, A. (1966) in: Solar Radio Observations during the IQSY, ed. by L. Mollwo and W. Dittmar, Berlin
Krüger, A. (1967): On homologous bursts of continuous solar radio emission, XIIth IAU General Assembly, Praha 1967
Krüger, A. (1968): Dr. Habil.-Thesis, Univ. Rostock
Krüger, A. (1969) in. Annals of the IQSY, Vol. 3, ed. by A. C. Stickland, M.I.T. Press
Krüger, A., Bumba, V., Howard, R., and Kleczek, J. (1968): Bull. Astr. Inst. Czechoslovakia **19**, 180
Krüger, A. and Lauter, E. A. (ed.) (1969): Geod. Geophys. Veröff. RII, No. 10
Krüger, A., Entzian, G., and Taubenheim, J. (1969) in: Solar Flares and Space Research, ed. by C. de Jager and Z. Švestka, North Holland Publ. Amsterdam
Kundu, M. R. (1959): Ann. d'Astrophys. **22**, 1
Kundu, M. R. (1961a): J. Geophys. Res. **66**, 4308
Kundu, M. R. (1961b): Astrophys. J. **134**, 96
Kundu, M. R. (1963): Space Sci. Rev. **2**, 438
Kundu, M. R. (1964): Solar Radio Astronomy, Vol. I and II, Report No. 64–4, Univ. Michigan Radio Astron. Obs. and Interscience Publ., New York (1965).
Kundu, M. R. (1965) in: The Solar Spectrum, ed. by C. de Jager, Reidel Publ. Dordrecht

Kundu, M. R., Roberts, J. A., Spencer, C. L. and Kuiper, J. W. (1961): Astrophys. J. **133**, 255
Kundu, M. R. and Spencer, C. L. (1963): Astrophys. J. **137**, 572
Künzel, H. (1960): Astron. Nachr. **285**, 271
Künzel, H. (1966): Astron. Nachr. **289**, 181

Landau, L. D. and Lifshits, E. M. (1961): The Classical Theory of Fields, Addison-Wesley, Reading, Mass.
Landauer, G. (1962): J. Nucl. Energy **C4**, 395
Lauter, E. A. (1961): Gerlands Beitr. Geophys. **71**, 58
Lauter, E. A., Entzian, G. and Knuth, R. (1960): Zs. f. Met. **14**, 275
Lauter, E. A. and Sprenger, K. (1958): Zs. f. Met. **12**, 205
Lehnert, B. (1964): Dynamics of Charged Particles, North-Holland Publ. Amsterdam
Lehnert, B. (1967): Plasma Physics **9**, 301
Leighton, R. B., Noyes, R. W., and Simon, G. W. (1962): Astrophys. J. **135**, 474
Le Roux, E. (1960): Ann. d'Astrophys. **23**, 1010
Le Roux, E. (1961): Ann. d'Astrophys. **24**, 71
Liemohn, H. B. (1965): Radio Sci. **69D**, 741
Lifshits, M. A. and Tsytovich, V. N. (1967): Zhurn. exp. teoret. fiz. **53**, 1610
Lin, R. P. (1968): Can. J. Phys. **46**, S757
Lin, R. P. (1970): Solar Physics **12**, 266 and 15, 453
Lingenfelter, R. E. and Ramanaty, R. (1967): Planet. Space Sci. **15**, 1303
Lingenfelter, R. E., Flamm, E. J., Canfield, E. H., and Kellman, S. (1965): J. Geophys. Res. **70**, 4077 and 4087
Linhart, J. C. (1961): Plasma Physics, North-Holland Publ. Amsterdam
Lucke, O. and Wagner, C.-U. (1961): Zs. f. Geophys. **27**, 257

Malville, J. M. and Smith, S. F. (1963): J. Geophys. Res. **68**, 3181
Mandelstam, S. L. (1965): Space Sci. Rev. **4**, 587
Marshall, L. (1956): Astrophys. J. **124**, 469
Martres, M. J., Michard, R., and Soru-Iscovici, I. (1966): Ann. d'Astrophys. **29**, 245
Maxwell, A. (1963): Planet. Space Sci. **11**, 897
Maxwell, A. (1965) in: The Solar Spectrum ed. by C. de Jager, Reidel Publ. Dordrecht
Maxwell, A., Hughes, M. P., and Thompson, A. R. (1963): J. Geophys. Res. **68**, 1347
Maxwell, A., Defouw, R. J., and Cummings, P. (1964): Planet. Space Sci. **12**, 435
McKenzie, J. F. (1963): Phil. Trans. Roy. Soc. London, **A255**, 585
McKenzie, J. F. (1966): Proc. Phys. Soc. London **87**, 349
McKenzie, J. F. (1967): Proc. Phys. Soc. London **91**, 532 and 537
Milne, E. A. (1921): Month. Not. Roy. Astr. Soc. **81**, 361
Minnaert, M. (1960) in: Radioastronomia solare, Varenna Summer School, Corso XII, Bologna
Minnaert, M. (1965) in: The Solar Spectrum, ed. by C. de Jager, Reidel Publ., Dordrecht
Miroshnichenko, L. I. (1968): Geomagn. i Aeronomiya **8**, 563
Mogilevsky, E. I. and Akinyan, S. T. (1961): Geomagn. i Aeronom. **1**, 921
Molchanov, A. P. (1961): Astron. Zhurn. **38**, 849
Molchanov, A. P. (1962): Solnechnye Dannye, (1962) No. **2**, 53
Mollwo, L. (1965): Beitr. Plasmaphys. **5**, 123
Mollwo, L. (1969): Private communication

Mollwo, L. (1970): Solar Physics, 12, 125
Moreton, G. E. (1965) in: Proc. Plasma Space Symp. ed. by C. C. Chang and S. S. Huang, Reidel Publ. Dordrecht
Moreton, G. E. and Severny, A. B. (1968): Solar Physics **3**, 282
Montgomery, D. L. and Tidman, D. A. (1964): Plasma Kinetic Theory, McGraw-Hill, New York
Mustel, E. R. (1955): Astron. Zhurn. **32**, 177
Mustel, E. R. (1966): Astron. Zhurn. **43**, 1121

Neupert, W. M. (1968): Astrophys. J. **153**, L59
Neupert, W. M. (1969): Ann. Rev. Astron. Astrophys. **7**, 121
Newkirk, G. (1961): Astrophys. J. **133**, 983
Newkirk, G. (1967): Ann. Rev. Astr. Astrophys. **5**, 213

Obayashi, T. (1962): J. Geophys. Res. **67**, 1717
Oster, L. (1959): Z. Astrophys. **47**, 169
Oster, L. (1961): Rev. Mod. Phys. **33**, 525 and Astrophys. J. **134**, 1010
Oster, L. (1963): Astrophys. J. **137**, 322 and **138**, 761
Oster, L. (1964): Phys. Fluids **7**, 263
Oster, L. and Sofia, S. (1966): Astrophys. J. **143**, 944
Oort, J. H. and Walraven, T. (1956): Bull. Astr. Inst. Netherlands, **12**, 285

Papas, C. H. (1965): Theory of Electromagnetic Wave Propagation, McGraw-Hill, New York
Parker, E. N. (1957): Phys. Rev. **107**, 830
Parker, E. N. (1958): Astrophys. J. **128**, 667 and Phys. Rev. **109**, 1328; **112**, 1429
Parker, E. N. (1963a): Astrophys. J., Suppl. **8**, 177
Parker, E. N. (1963b): Interplanetary Dynamical Processes, Interscience Publ. New York
Parker, E. N. (1964): Research in Geophysics, ed. by H. Odishaw, Vol. 1, M.I.T. Press
Parker, E. N. and Tidman, D. A. (1958): Phys. Rev. **111**, 1206 and **112**, 1048
Petchek, H. E. (1964) in: The Physics of Solar Flares, ed. by N. Hess, Washington
Peterson, A. M. and Hower, G. L. (1963): J. Geophys. Res. **68**, 723
Peterson, L. E. and Winckler, J. R. (1959): J. Geophys. Res. **64**, 697
Pick-Gutman, M. (1961): Ann. d'Astrophys. **24**, 183
Piddington, J. H. (1955): Phil. Mag. **46**, 1037
Piddington, J. H. (1967): Planet. Space Sci. **15**, 733
Piddington, J. H. and Minnett, H. C. (1951): Aust. J. Sci. Res. **A4**, 131
Pintér, Š. (1969): Solar Physics **8**, 149
Pneuman, G. W. (1967): Solar Physics **2**, 462
Pounds, K. A. (1965): Ann. d'Astrophys. **28**, 132
Priester, W. and Rosenberg, J. (1965): NASA Techn. Note D-2888

Rao, U. V. G. (1970): Solar Physics **14**, 389
Ramanaty, R. (1968): J. Geophys. Res. **73**, 3573
Ramanaty, R. and Lingenfelter, R. E. (1966): J. Geophys. Res. **71**, 3687
Ramanaty, R. and Lingenfelter, R. E. (1967): J. Geophys. Res. **72**, 879
Ratcliffe, J. A. (1959): The Magneto-Ionic Theory and its Application to the Ionosphere, Univ. Press. Cambridge

Rawer. K. and Suchy, K. (1967) in: Handbuch d. Physik, Herausg. S. Flügge, Bd. 49, Geophysik III/2, Springer Verlag Berlin, Heidelberg, New York
Razin, V. A. (1960): Izvest. vys. ucheb. zav., Radiofizika **3**, 584 and 921
Roederer, J. G. (1964): Space Sci. Rev. **3**, 487
Roberts, J. A. (1959): Aust. J. Phys. **12**, 327
Rydbeck, O. R. (1944): On the Propagation of Radio Waves, Trans. Chalmers Univ. of Technol., No. 34

Sagdeev, R. Z. and Shafranov, V. D. (1960): Zhurn. exp. teoret. fiz. **39**, 181
Sandström, A. E. (1965): Cosmic Ray Physics, North-Holland Publ., Amsterdam
Sakurai, K. (1961): J. Geomagn. Geoelectr. **12**, 70
Sakurai, K. (1965): Publ. Astron. Soc. Japan **17**, 403
Sakurai, K. (1966a): Rep. Ionos. Space Res. Japan **20**, 233
Sakurai, K. (1966b): Publ. Astron. Soc. Japan **18**, 77
Sazonov, V. N. and Tsytovich, V. N. (1968): Izv. vys. ucheb. zav. radiofizika **11**, 1287
Sazonov, V. N. (1969a): Astron. Zhurn. **46**, 502
Sazonov, V. N. (1969b): Zhurn. exp. teoret. fiz. **56**, 1075
Scarf, F. L., Bernstein, W. and Fredricks, R. W. (1965): J. Geophys. Res. **70**, 9
Schatzmann, E. (1965) in: The Solar Spectrum, ed. by C. de Jager, Reidel Publ. Dordrecht
Schatzmann, E. (1967): Solar Physics **1**, 411
Scheuer, P. A. G. (1960): Mon. Not. Roy. Astr. Soc. **120**, 231
Scheuer, P. A. G. (1968): Astrophys. J. **151**, L 139
Schott, G. A. (1912): Electromagnetic Radiation, Univ. Press Cambridge
Schwarzschild, K. (1914): Sitz. Ber. Preuß. Akad. Wiss. 1183
Schwinger, J. (1949): Phys. Rev. **75**, 1912
Seshadri, S. R. and Tuan, H. S. (1965): Radio Science **69D**, 767
Severny, A. B. (1958): Izv. Krymsk. Astrofiz. Obs. **20**, 22
Severny, A. B. (1961): Astron. Zhurn. **38**, 402
Severny, A. B. (1964): Ann. Rev. Astr. Astrophys. **2**, 363 and Space Sci. Rev. **3**, 451
Severny, A. B. (1965) in: The Solar System, ed. by C. de Jager, Reidel Publ., Dordrecht
Severny, A. B. (1966): Astron. Zhurn. **43**, 465
Severny, A. B. (1969): Solar Flares and Space Research ed. by C. de Jager and Z. Švestka, North-Holland Publ. Amsterdam
Severny, A. B. and Shabansky, V. P. (1961): Izv. Krymsk. Astrofiz. Obs. **25**, 88
Severny, A. B. and Steshenko, N. V. (1969): Solnechno-Zenmaya Fizika (Moscow) **1**, 3
Shklovsky, I. S. (1965): Physics of the Solar Corona, Pergamon Press
Simon, M. (1969a): Astrophys. J. **156**, 341
Simon, M. (1969b): Astrophys. Letters **3**, 23
Simon, P. (1960): Ann. d'Astrophys. **23**, 102
Simon, P. (1969): I.U.W.D.S. Synoptic Codes for Solor and Geophysical Data, p. 7
Sitenko, A. G. and Stepanov, K. N. (1956): Zhurn. exp. teoret. fiz. **31**, 642
Slonim, Y. M. (1968): Astron. Zhurn. **45**, 286 and 726
Slonim, Y. M. (1969): Astron. Zhurn. **46**, 570 and 697
Smerd, S. F. (1964) in: Research in Geophysics, ed. by H. Odishaw, Vol. 1, M.I.T. Press
Smerd, S. F. (1965) in: The Solar Spectrum, ed. by C. de Jager, Reidel Publ., Dordrecht
Smith, H. J. and Smith, E. van P. (1963): Solar Flares, MacMillan, New York
Snyder, W. and Helliwell, R. A. (1952): J. Geophys. Res. **57**, 73
Sokolov, A. A. and Ternov, J. M. (1966): Synchrotron-Radiation (in russ.), 'Nauka' Moscow

Sokolov, A. A., Ternov, I. M. , Bagrov, V. G., Galtsov, D. V., and Shukovsky, V. T. (1968): Z. Physik **211**, 1
Sollfrey, W. and Yura, H. T. (1965): Phys. Rev. **139**, A48
Spitzer, L. (1962): Physics of Fully Ionized Gases, 2nd ed., Interscience Publ., New York
Steenbeck, M., Krause, K. and Rädler, K.-H. (1966): Z. Naturforsch. **21a**, 369 and 1285
Stein, W. A. and Ney, E. P. (1963): J. Geophys. Res. **68**, 65
Stepanov, K. N. (1958): Zhurn. exp. teoret. fiz. **35**, 1155 and 283
Stepanov, K. N. (1959): Zhurn. exp. teoret. fiz. **36**, 1457
Stepanov, K. N. and Pakhomov, V. I. (1960): Zhurn. exp. teoret. fiz. **38**, 1564
Stepanyan, A. A. and Vladimirsky, B. M. (1961): Astron. Zhurn. **38**, 439
Stern, D. P. (1966): Space Sci. Rev. **6**, 147
Steward, R. T. (1965): Austr. J. Phys. **18**, 67
Stickland, A. C. (ed.) (1969): Annals of the IQSY, Vol. III The Proton Flare Project, Vol. IV Solar-Terrestrial Physics, M.I.T. Press
Stix, T. H. (1962): The Theory of Plasma Waves, McGraw-Hill, New York
Straka, R. M. and Barron, W. R. (1969): Paper presented at XV Electromagnetic Wave Propagation Meeting of AGARD, St. Jovite, Canada
Sturrock, P. A. (1964a): Nature, London **203**, 285
Sturrock, P. A. (1964b) in: The Physics of Solar Flares ed. by W. N. Hess, Washington
Sturrock, P. A. (1966): Nature **211**, 695
Sturrock, P. A. (1968) in: Structure and Development of Solar Active Regions, IAU- Symp. No. 35, ed. by K. O. Kiepenheuer, Reidel Publ. Dordrecht
Sturrock, P. A. and Coppi, B. (1964): Nature, London **204**, 4953
Suemoto, Z. and Hiei, E. (1959): Publ. Astr. Soc. Japan **11**, 152
Suemoto, Z., Hiei, E., and Hirayama, T. (1962): J. Phys. Soc. Japan, **17**, Suppl. A-II, 231
Švestka, Z. (1966): Space Sci. Rev. **5**, 388 and Bull. Astr. Inst. Czechoslovakia **17**, 262
Švestka, Z. (1967): Bull. Astr. Inst. Czechoslovakia **18**, 55
Švestka, Z. (1968): Solar Physics **4**, 18
Švestka, Z. and Olmr, J. (1966): Bull. Astr. Inst. Czechoslovakia **17**, 4
Švestka, Z. and Simon, P. (1969): Solar Physics **10**, 3
Swann, W. F. G. (1933): Phys. Rev. **43**, 217 and **44**, 224
Sweet, P. A. (1958) in: Electromagnetic Phenomena in Cosmical Physics, IAU-Symposium No. 6, ed. by B. Lehnert, Univ. Press, Cambridge
Sweet, P. A. (1965) in: Stellar and Solar Magnetic Fields, IAU-Symposium No. 22, ed. by R. Lüst, North-Holland Publ., Amsterdam
Sweet, P. A. (1969): Ann. Rev. Astron. Astrophys. **7**, 140
Syrovatsky, S. I. (1957): Usp. Fiz. Nauk **62**, 247
Syrovatsky, S. I. (1966): Astron. Zhurn. **43**, 2 and Zhurn. exp. teoret. fiz. **50**, 1133
Syrovatsky, S. I. (1967) in: Radio Astronomy and the Galactic System, IAU-Symposium No. **31**, ed. by H. van Woerden, North-Holland Publ., Amsterdam
Syrovatsky, S. I. (1968): Zhurn. exp. teoret. fiz. **54**, 1422
Syrovatsky, S. I. (1969) in: Solar Flares and Space Research ed. by C. de Jager and Z. Švestka, North-Holland Publ., Amsterdam

Takakura, T. (1960): Publ. Astr. Soc. Japan **12**, 55 and **325** and 352
Takakura, T. (1961): Publ. Astr. Soc. Japan **13**, 166 and 312
Takakura, T. (1962): J. Phys. Soc. Japan **17**, Suppl. A-II, 243
Takakura, T. (1963): Publ. Astr. Soc. Japan **15**, 327 and 462

Takakura, T. (1964): Publ. Astr. Soc. Japan **16**, 230
Takakura, T. (1966): Space Sci. Rev. **5**, 80
Takakura, T. (1967): Solar Physics **1**, 304
Takakura, T. (1968): Solar Physics **6**, 133
Takakura, T. (1969): Solar Physics **6**, 133
Takakura, T. and Kai, K. (1961): Publ. Astron. Soc. Japan **13**, 94
Takakura, T. and Kai, K. (1966): Publ. Astr. Soc. Japan 18, 57
Takakura, T. and Ono, M. (1962): J. Phys. Soc. Japan **17**, Suppl. A-II, 207
Takakura, T., Uchida, Y., and Kai, K. (1968): Solar Physics **4**, 45
Takakura, T. and Uchida, Y. (1968): Astrophys. Letters **1**, 147 and **2**, 87
Tanaka, H. and Enomé, S. (1970): Nature **225**, 435
Tanaka, H. and Kakinuma, T. (1959) in: Paris Symposium on Radio Astronomy, ed. by R. N. Bracewell Stanford Univ. Press.
Tanaka, H. and Kakinuma, T. (1960): Proc. Res. Inst. Atmos. Nagoya Univ. **7**, 72
Tanaka, H. and Kakinuma, T. (1961): Proc. Res. Inst. Atmos. Nagoya Univ. **8**, 39
Tanaka, H. and Kakinuma, T. (1962): J. Phys. Soc. Japan **17**, Suppl. A-II, 211
Tanaka, H. and Kakinuma, T. (1964): Rep. Ionos. Space Res. Japan **18**, 32
Tandberg-Hanssen, E. (1967): Solar Activity, Blaisdell Publ. Co. Waltham/Mass.
Teske, R. G. and Thomas, R. J. (1969): Solar Physics **8**, 348
Thomas, R. N. and Athey, R. G. (1961): Physics of the Solar Chromosphere, Interscience Publ., New York
Thompson, A. R. and Maxwell, A. (1962): Astrophys. J. **136**, 546
Tidman, D. A. (1965): Planet. Space Sci. **13**, 781
Tidman, D. A., Birmingham, T. J., and Stainer, H. M. (1966): Astrophys. J. **146**, 207
Tidman, D. A. and Dupree, T. H. (1965): Phys. Fluids **8**, 1860
Tlamicha, A., Olmr, J., Fürstenberg, F., and Krüger, A. (1966): Mber. Dt. Akad. Wiss. Berlin **8**, 803
Tlamicha, A., Olmr, J., Fürstenberg, F., and Krüger, A. (1967): Bull. Astr. Inst. Czechoslovakia **18**, 70
Tousey, R. (1963): Space Sci. Rev. **2**, 3
Trubnikov, B. A. (1958): Dokl. Akad. Nauk SSSR **118**, 913
Trubnikov, B. A. and Bashanova, A. F. (1959): Phys. Abhandl. SU., Neue Folge **1**, 472
Tsuchiya, A. (1968): Solar Physics **3**, 346
Tsytovich, V. N. (1951): Vestnik Mos. Gos. Univ. **11**, 27
Tsytovich, V. N. (1967): Nonlinear Effects in Plasma (in russ.), 'Nauka', Moscow
Twiss, R. Q. (1954): Phil. Mag. **45**, 249
Twiss, R. Q. (1958): Aust. J. Phys. **11**, 564
Twiss, R. Q. (1962): Astrophys. J. **136**, 438
Twiss, R. Q. and Roberts, J. A. (1958): Aust. J. Physics **11**, 424

Uchida, Y. (1968): Solar Physics **4**, 30
Unsöld, A. (1955): Physik der Sternatmosphären, Springer-Verlag, Berlin, Göttingen, Heidelberg
Unsöld, A. (1960): Z. Astrophys. **50**, 48 and 57

Vjalshin, G. F., Karaev, A., Krat, V. A., Křivsky, L., Krüger, A., Künzel, H., and Mamedov, S. (1967): Solnechnye Dannye 1967, No. **3**, 90
Vladimirsky, V. V. (1948): Zhurn. exp. teoret. fiz. **18**, 393

Waldmeier, M. (1963): Z. Astrophys. **58**, 77

Waldmeier, A. and Müller, H. (1950): Z. Astrophys. 27, 58
Wallis, G. (1959) in: Paris Symposium on Radio Astronomy, ed. by R. N. Bracewell, Stanford Univ. Press
Wallis, G. (1962): Dr. Habil. Thesis, Humboldt-Univ. Berlin
Warwick, C. S. (1965): Astrophys. J. 141, 500
Warwick, C. S. (1966): Astrophys. J. 145, 215
Warwick, C. S. and Wood-Haurwitz, M. (1962): J. Geophys. Res. 67, 1317
Warwick, J. W. (1962): Publ. Astr. Soc. Pacific 74, 302
Warwick, J. W. (1957): Astrophys. J. 125, 811
Warwick, J. W. (1965) in: Solar System Radio Astronomy, ed. by J. Aarons, Plenum Press, New York
Warwick, J. W. (1967): Astrophys. J. 150, 1081
Wentzel, D. G. (1963): Astrophys. J. 137, 135
Wentzel, D. G. (1964) in: The Physics of Solar Flares, ed. by W. N. Hess, Washington, and Astrophys. J. 140, 1563
Wentzel, D. G. (1965): J. Geophys. Res. 70, 2716
Weiss, A. A. (1963): Aust. J. Phys. 16, 240 and 526
Weiss, A. A. (1965): Aust. J. Phys. 18, 143
Westfold, K. C. (1957): Phil. Mag. 2, 1287
Westfold, K. C. (1959): Astrophys. J. 130, 241
Wilcox, J. M. and Schatten, K. H. (1967): Astrophys. J. 147, 364
Wild, J. P. (1962): J. Phys. Soc. Japan 17, Suppl. A-II, 249
Wild, J. P. (1963) in: The Solar Corona, I.A.U.-Symposium No. 16, ed. by J. W. Evans, Acad. Press New York, London
Wild, J. P. (1963) in: Radio Astronomy Today, ed. by H. P. Palmer, R. D. Davies, and M. I. Large, Manchester
Wild, J. P. (1969): Proc. Astron. Soc. Australia 1, 181
Wild, J. P. and McCready, L. L. (1950): Austr. J. Sci. Res. (A) 3, 387
Wild, J. P., Sheridan, K. V., and Trent, G. H. (1959) in: Paris Symposium on Radio Astronomy, ed. by R. N. Bracewell, Stanford Univ. Press
Wild, J. P., Smerd, S. F., and Weiss, A. A. (1963): Ann. Rev. Astr. Astrophys. 1, 291
Xanthakis, J. N. ed. (1967): Solar Physics, Interscience Publ. New York
Yip, W. K. (1967): Aust. J. Phys. 20, 421
Yurovsky, Yu. F. (1969): Solnechno-Zemnaya Fizika (Moscow) 1, 15
Zajtsev, V. V. and Kaplan, S. A. (1966): Akad. Nauk Armen. SSR, Astrofizika 2, 169
Zheleznyakov, V. V. (1962): Astron. Zhurn. 39, 5
Zheleznyakov, V. V. (1963): Astron. Zhurn. 40, 829
Zheleznyakov, V. V. (1964a): Izvest. vys. ucheb. zav. Radiofizika 7, 67
Zheleznyakov, V. V. (1964): Radio Emission from the Sun and the Planets (in russ.), Moscow (1970 – engl. transl. Pergamon Press)
Zheleznyakov, V. V. (1965): Astron. Zhurn. 42, 96
Zheleznyakov, V. V. (1966): Zhurn. exp. teoret. fiz. 51, 570 and Izvest. vys. ucheb. zav., Radiofizika 9, 1058
Zheleznyakov, V. V. (1967): Astron. Zhurn. 44, 42
Zheleznyakov, V. V. (1968): Astrophys. Space Sci. 2, 417
Zheleznyakov, V. V. and Trachtengerz, V. Y. (1965): Astron. Zhurn. 42, 1005
Zheleznyakov, V. V. and Zajtsev, V. V. (1968): Astron. Zhurn. 45, 19
Zheleznyakov, V. V. and Zlotnik, E. Y. (1962): Izvest. vys. ucheb. zav., Radiofizika 5, 644
Zirin, H. (1966): The Solar Atmosphere, Blaisdell Publ. Co. Waltham/Mass.

Appendix 1

List of major type IV events 1956—1969

Explanations

Column

1 – Date

2–3 – The start times and durations contain only approximate informations with special respect to the microwave region mainly for the purpose of identification of the events.

4 – The 3 cm maximum flux values are subject to uncertainties of the absolute calibration (~ ± 20 per cent in the period before 1964) and some other errors by missing standardization of the measurements of different stations. The flux values are mostly rounded up to 10 s. u. (The fluxes are taken as peak values minus pre-burst level).

5 – Because of the incompleteness of the observations the frequency extent in many cases indicates the observed spectral band but not necessarily the full actual emitted frequency range. Values in parentheses denote the band limit of spectrographic records.

6 – Due to the lack of more detailed observations (interferometric and polarization) the spectral classification provides only a rough estimation of the presence of certain type IV components. There has been made no difference between '*M*' and 'μ' bursts in the microwave region. Components which are not fully measured are given in parentheses. The symbol d^+ denotes exceptionally strong dm-components.

7–9 – Flare data

10–12 – Particle data (GLE – ground level enhancements, PCA – polar cap absorptions, SSE – satellite sensored events) are given by start times or crosses if the start times are not exactly known to us. Dates are attributed in parentheses if they differ from those given in column 1.

13 – MacMath-Hulbert plage numbers of the plage regions belonging to the related flare centres.

Note added in proof: The content of Appendix 1 and 2 is taken essentially from the „Catalogue of Solar Type IV Radio Events" by A. Krüger, F. Fürstenberg and K. H. Fricke, HHI Suppl. Ser. Solar Data Vol. II (1971) No. 5.

	Type IV Burst					Flare			Particles			Plage
Date	Start UT	Dur. h min	3 cm flux s.u.	Frequ. ext. MHz	Spectr.	Start. UT	Position	Imp.	GLE	PCA Start. UT	SSE Start. UT	Reg. No.
1956												
Feb. 16	17.56	6 24	-	167 2800	µdmAB	18.05	21N 08E	2+				3400
Feb. 17	11.10	0 35	440	81 9400	µmA	11.00	21N 04W	2+				3400
Feb. 19	14.20	1 05	-	81 2800	µdmAB	<14.30	23N 27W	2				3400
Feb. 23	03.34	0 51	-	200 3750	µmAB	<03.34	23N 80W	3		04.0		3400
Mar. 15	16.21	1 34	-	167 2800	µdmAB	<16.25	22N 22E	2+				3432
Aug. 29	09.37	0 11	120	536 9400	µd	09.37	30N 60W	3				3629
Aug. 31	12.31	1 29	>1225	81 9400	µdmAB	12.26	16N 15E	3		14.5		3643
Nov. 7	11.03	1 52	340	81 9400	µdmAB	11.09	17S 30E	3				3751
Nov. 13	14.26	0 24	30	169 9400	µmAB	<14.30	16N 10W	2		20.0		3753
Nov. 14	10.32	1 13	745	169 9400	µdmAB	10.37	20S 54W	3				3751
Nov. 19	08.57	1 48	170	81 9400	µmAB	<08.34	14S 42W	2				3755
Nov. 20	10.00	2 45	>5000	81 9400	µdmA	10.02	14S 56W	2+				3755
1957												
Jan. 6	17.02	1 58	-	167 2800	µB	no flare patrol						
Jan. 20	10.40	1 50	70	81 9400	µmA	<11.00	29S 18W	2+		15.0		3820
Apr. 3	08.26	1 04	370	200 9400	µmA	08.25	14S 60W	3		11.0		3907
Apr. 16	10.38	0 57	990	200 9400	µmA	10.40	30N 85E	2+				3491
Apr. 17	20.06	0 49	-	200 2800	µmA	no flare patrol						
June 3	10.42	1 48	350	81 9400	µdmAB	10.40	18S 15W	2+				3993
June 4	08.58	1 21	170	81 9400	µdmAB	08.59	17S 27W	2				3993
June 19	16.08	1 30	>800	100 9400	µdmAB	16.09	21N 40E	2+		22.3		4024
June 28	07.07	1 28	60	169 9500	µdmAB	06.58	10N 27E	2		04.0		4039
June 28	12.22	1 23	280	38 9400	µmAB	<12.23	12N 21E	2				4039
June 30	09.53	3 37	130	169 9400	µdmAB	09.24	10N 03W	2+				4039

Date	Type IV Burst Start. UT	Dur. h min	3 cm flux s.u.	Frequ. ext. MHz	Spectr.	Flare Start.	Position	Imp.	Particles GLE	PCA Start. UT	SSE Start. UT	Plage Reg. No.
1957												
July 3	07.22	0 58	2600	80 9500	µdmAB	07.12	14N 40W	3		08.3		4039
July 3	08.32	1 13		80 9500		08.30	09N 42W	2+				4039
July 16	17.40	0 35	460	169 9400	µmA	17.42	32S 30W	1+				4061
July 21	13.22	2 08	-	169 2800	µdmA	13.20	30N 11E	2				4073
July 21	21.45	1 30	-	167 9400	µmAB	21.34	30N 07E	1+				4073
July 22	06.17	1 03	90	169 9400	µmAB	<06.07	29N 03E	2				4073
						06.20	11N 53E	1+				4075
July 24	18.01	1 19	-	167 2980	µmA	17.12	24S 27W	2+		19.5		4070
Aug. 28	09.15	1 25	400	81 9400	µmA	08.10	30S 35E	3		13.0		4125
Aug. 30	22.10	0 30	-	167 9400	µmA	no flare patrol						
Aug. 31	05.48	3 02	700	169 9400	µdmAB	05.21	32S 04W	2+				4125
Aug. 31	12.58	2 52	>600	80 9400	µdmAB	12.57	25N 02W	3		13.7		4124
Sept. 2	12.55	3 05	30	81 9400	µdmA	12.28	34S 44W	2		15.0		4125
						12.57	10N 26W	1+				4124
Sept. 3	14.15	1 22	> 220	169 9400	µdmAB	14.12	25N 30W	3		15.0		4124
Sept.11	02.35	8 45	140	169 9500	µdmAB	<02.36	13N 02W	3				4134
Sept.12	15.14	2 46	850	81 9400	µdmAB	15.10	10N 18W	2		16.0		4134
Sept.13	14.11	2 19	240	169 9400	µdmA	13.13	15S 26W	2				4138
						14.10	09N 32W	1+				4134
Sept.18	18.05	> 4 35	-	167 2800	µdmAB	<16.58	23N 08E	3		20.0		4151
Sept.21	13.30	0 20	760	81 9400	µmAB	13.30	11N 07W	3		11.0 15.0		4152
Sept.22	12.51	0 49	630	169 9400	µdmA	12.48	07N 35W	2+		10.0		4152
Sept.26	19.15	5 15	-	167 2800	µmA	19.07	24N 16E	3		21.0		4159
Oct. 20	16.44	5 26	-	167 2800	µdmAB	16.37	27S 38W	3+		17.0 21.0		4189

Nov. 05	12.04	0 21	400	81 9400	dmA	< 12.05	24S 54W	2		4207
Nov. 24	08.57	1 18	240	169 9400	µdmAB	08.48	14S 37E	3	02.0	4263
Nov. 24	18.05	1 20	-	200 2800	µmA	16.11	14S 23E	1		4263
Dec. 14	12.28	1 07	670	169 9400	µmA	< 12.45	18N 77E	2+	x	4317
Dec. 16	11.34	1 11	20	81 9400	µdmA	11.25	16N 51E	2+		4317
Dec. 17	07.36	0 54	80	169 9400	µmA	< 07.34	20N 41E	2+	03.0	4317
Dec. 17	10.20	1 05	-	81 1500	dmAB					
Dec. 19	09.00	0 45	80	169 9400	µmAB	< 07.57	20N 14E	2+		4317
Dec. 23	14.38	0 32	-	169 2980	µmA	14.36	18N 45W	1+		4314
Dec. 28	22.28	0 27	> 820	100 9500	µdmAB	22.29	25N 50W	2		4321
1958										
Jan. 07	18.22	2 08	-	167 2980	µdmA	{18.20	18S 42E	2		4355
						{18.55	12N 03W	1		4347
Jan. 20	14.45	0 22	50	81 9400	µdmA	14.35	27N 44W	2+		4381
Jan. 25	12.04	2 56	70	169 9400	µdmA	12.05	19S 11E	2	16.0	4382
Feb. 09	21.12	2 13	-	167 1420	dmAB	21.08	13S 14W	2	(10.)05.0	4400
Mar. 01	09.11	0 48	680	169 9400	µdmAB	09.05	10S 46W	3		4436
Mar. 03	10.10	4 00	870	80 9400	µdmAB	< 10.05	17S 60E	3		4445
Mar. 14	14.53	0 32	50	81 9400	µdmA	14.54	21S 85W	2	15.0	4445
Mar. 23	09.55	2 20	-	80 3000	µmAB	09.47	14S 80E	3+	15.0	4476
Mar. 30	09.09	1 16	70	23 9800	µdmAB	< 09.15	15S 20W	2+	(31.)16.0	4476
June 04	21.39	0 31	> 670	167 9500	µmAB	21.47	15N 58W	1+	23.0	4578
June 05	08.40	0 20	770	169 9400	µmA	08.35	15N 63W	2		4578
June 05	16.55	0 40	100	167 9400	µmA	16.56	21S 70E	2+	06.0	4598
June 06	04.33	0 42	> 710	169 9500	µmA	04.36	16N 78W	2	13.7	4578
July 07	00.27	1 53	1480	200 9500	µmA	00.20	26N 08W	3+	01.0	4630
July 11	07.51	1 09	100	178 9500	µmAB	07.40	25S 28E	2+		4642
July 23	13.22	0 33	-	167 2800	µmAB	12.59	16N 35E	2+		4657
July 29	02.44	5 26	5410	169 9500	µdmAB	< 02.59	14S 44W	3	04.0	4659
Aug. 16	04.34	2 31	6910	169 9500	µdmAB	< 04.32	14S 49W	3+	06.0	4686
Aug. 18	08.10	2 10	240	23 9500	µdmA	07.47	19N 50E	2		4708
Aug. 19	21.50	2 20	340	167 9500	µdmAB	21.18	18N 26E	2		4708

	Type IV Burst					Flare			Particles			Plage
Date	Start. UT	Dur. h min	3 cm flux s.u.	Frequ. ext. MHz	Spectr.	Start. UT	Position	Imp.	GLE	PCA Start. UT	SSE Start. UT	Reg. No.
1958												
Aug. 22	14.23	1 47	440	167 9400	µmAB	14.17	18N 08W	3		15.13		4708
Aug. 26	00.17	1 53	5440	167 9500	µdmAB	00.05	20N 54W	3		01.0		4708
Aug. 28	10.09	1 20	600	81 9400	µdmAB	10.25	16S 67W	2+				4712
Sept.14	08.32	1 03	890	81 9400	µmA	08.22	10S 80W	2+		10.0		4741
Sept.22	07.35	0 30	10	207 9400	µmAB	07.38	20S 42W	2		14.0		4759
Oct. 21	23.23	1 12	3120	167 9500	µmAB	23.18	04S 22W	2+				4826
Oct. 24	14.39	0 41	50	81 9400	µmA	14.10	04S 57W	2+				4826
Nov. 24	16.00	1 50	-	167 2800	µmAB	16.07	12S 09W	3				4883
Dec. 11	14.26	0 12	-	169 2980	µmAB	14.30	02S 03E	1				4913
Dec. 11	18.04	0 21	-	167 2800	µmAB	{18.02	02S 00	2				4913
						19.30}	02S 03W	1+				4913
Dec. 12	12.51	3 59	780	81 9400	µdmAB	12.15	03S 08W	2+				4913
Dec. 23	05.35	1 40	1690	169 9500	µdmA	< 05.45	15S 66E	2				4934
Dec. 31	16.57	15	-	167 2800	µdmA	< 16.56	18S 54W	3				4934
1959												
Jan. 26	10.30	0 40	290	210 9400	µmAB	10.27	15N 60W	2+		14.0		4969
Feb. 08	13.42	0 11	30	169 9400	µdmB	13.43	10N 90E	1				5009
Feb. 09	01.58	0 17	510	200 9500	µmAB	00.58	10N 90E	2				5009
Feb. 09	09.59	0 26	20	169 9400	µmAB	< 07.35	09N 87E	2+				5009
Feb. 09	13.10	1 05	80	200 3000	µmA	12.30	09N 88E	2				5009
Feb. 12	22.57	0 48	400	167 9500	µmA	< 23.01	13N 48E	3		(13.)09.0		5009
Mar. 24	10.00	2 58	780	111 9400	µdmB	09.58	29N 77W	3		(25.)03.0		5054
Apr. 08	09.14	0 41	660	169 9400	µmA	09.03	27N 85E	2+				5093
May 08	22.54	25	>1160	167 9500	µdmA	22.52	21N 83E	2+				5148
May 10	21.00	2 45	2440	167 9500	µmAB	20.55	20N 47E	3+		24.00		5148
May 11	20.10	0 35	-	167 2800	µmA	20.06	10N 40E	2+				5148
May 13	05.10	0 20	1240	127 9500	µmAB	05.09	21N 25E	2+		08.00		5148

June 09	04.50	3 10	-	127	3750	µdmAB	04.30	31S	20E	1			5194
June 09	16.35	1 59	430	167	9400	µmA	< 18.00	17N	90E	2			5204
June 18	10.55	2 43	870	545	19000	µd	> 11.05	08N	61E	2			5219
							11.30	16N	12W	3+			5204
July 09	20.20	2 20	-	167	9400	µdmAB	19.30	19N	67E	2		20.0	5265
							< 21.15	20N	51E	1			5265
July 10	02.08	1 37	13500	167	9500	µdmAB	02.06	20N	63E	3+		04.0	5265
July 14	03.30	3 25	3120	200	9500	µdmAB	< 03.25	17N	05E	3+		04.7	5265
July 16	21.14	2 46	-	167	9400	µdmAB	21.14	14N	30W	3	x	22.8	5265
July 27	12.25	45	160	18	9400	µdmA	12.24	14N	50E	2			5294
Aug. 14	01.30	1 10	-	200	9400	µmA	00.40	12N	30E	2			5323
Aug. 18	10.25	1 15	680	178	19000	µdmAB	10.14	12N	33W	3		10.7	5323
Aug. 28	00.27	2 25	570	18	9400	µdmAB	00.27	12N	72E	1+			5344
Aug. 31	18.47	0 33	-	200	2800	µmA	18.50	10N	11E	1+			5344
Sept.01	19.28	0 32	-	545	2800	µmA	19.23	12N	60E	2+		(02.)04.0	5354
Nov. 30	02.47	2 13	3800	100	9400	µmA	02.47	08N	16E	2+			5476
Nov. 30	17.38	6 00	-	167	2800	µmA	17.20	08N	06E	3			5476
Dec. 02	12.46	0 49	900	167	9400	µmAB	12.19	07N	16W	2+			5476
Dec. 21	00.43	0 37	33C	208	9500	µmA	00.43	03S	55W	1+			5494
1960													
Jan. 11	20.56	0 44	-	18	2800	µmAB	< 20.40	22N	02E	3		22.0	5527
Jan. 15	13.35	0 30	210	200	9400	µdmA	13.36	20S	69W	2			5525
Feb. 22	13.53	0 23	455	200	9400	µmA	13.52	09N	41E	2+			5581
Mar. 28	20.48	5 00	-	18	2800	µdmA	20.42	14N	37E	2			5615
Mar. 29	06.56	2 14	7060	200	9500	µdmAB	< 06.40	12N	30E	2+		08.0	5615
Mar. 30	02.15	3 00	130	545	9500	µdmA	02.16	09N	15E	1+		09.5	5615
Mar. 30	15.21	2 54	7300	167	9400	µmAB	14.55	12N	12E	2		15.0	5615
Apr. 01	08.45	1 50	8000 ?	200	9400	µmA	08.43	12N	12W	3		09.5	5615
Apr. 05	01.24	2 21	8800	200	9400	µdmA	< 02.15	12N	62W	2+		04.0	5615
Apr. 28	01.17	0 23	210	1000	9500	µmA	< 01.30	05S	35E	3		03.0	5645
Apr. 29	01.40	1 20	-	545	9400	µmA	< 01.07	14N	21W	2+		02.0	5642
May 04	10.12	2 19	380	23	9400	µmA	10.00	13N	90W	2	10.5		5642
May 06	14.08	2 12	50	231	9400	µdmA	14.04	08S	08E	3		16.0	5653

Date	Type IV Burst Start. UT	Dur. h min	3 cm flux s.u.	Frequ. ext. MHz	Spectr.	Flare Start. UT	Flare Position	Flare Imp.	Particles GLE	Particles PCA Start. UT	Particles SSB Start. UT	Plage Reg. No.
1960												
May 12	13.40	2 42	200	16 9400	µmAB	< 13.42	30N 59W	1+				5654
May 13	05.17	2 03	9990	200 9400	µdmA	05.12	29N 68W	3		06.2		5654
May 26	09.02	0 48	> 1150	200 9400	µmA	08.18	14N 15W	2		10.0		5669
June 01	08.34	0 51	> 3400	23 9400	µmA	08.24	29N 47E	3+		14.0		5680
June 25	11.53	3 07	300	200 9100	µdmA	< 11.31	20N 06E	3				5713
June 25	16.59	1 41	-	18 2800	µmAB	16.59	19N 01W	1		17.0		5713
June 25	20.37	0 48	-	167 2800	µmAB	20.39	19N 04W	2+				5713
June 27	00.02	0 48	50	200 9400	µmA	23.58	08S 34E	3		23.0		5719
June 27	04.22	0 58	3300	200 9400	µmAB	04.18	21N 19W	1+				5713
June 27	21.40	1 25	105	18 9400	µdmAB	21.40	21N 25W	2+				5713
June 29	01.36	0 19	1450	200 9400	µmA	01.25	20N 50W	1				5713
June 30	10.23	0 42	210	536 9400	µmA	< 10.29	18N 64W	1				5713
July 14	10.57	0 13	330	808 9400	µd	10.56	15N 52W	2				5740
Aug. 11	19.23	0 20	-	18 2800	µmA	19.16	22N 27E	3		24.0		5794
Sept.03	00.35	1 30	14700	200 9500 }	µdmA	00.37	19N 88E	2+	02.5			5837
Sept.03	02.16	3 25	30	545 9500 }						05.0		
Sept.16	17.10	1 15	740	18 9100 }	µdmA	17.10	22S 68E	1				5858
Sept.16	18.40	0 25	-	200 2800 }								
Sept.26	05.30	1 00	2600	200 9500	µdmA	05.25	19S 64W	2		07.0		5858
Oct. 09	11.57	0 48	40	231 9400	µmA	11.50	16S 13W	1+				5880
Oct. 10	06.54	0 41	280	231 9400	µmA	< 07.10	17S 23W	1+				5880
Oct. 11	05.20	0 40	2600	200 9500	µmA	05.17	16S 36W	2				5880
Oct. 29	10.25	1 05	820	23 9400	µmA	10.26	22N 26E	3		12.0		5909
Nov. 10	10.12	1 33	320	127 9400	µdmAB	10.09	28N 25E	3		18.0		5925
Nov. 10	12.17	0 23	-	127 1500								
Nov. 11	03.15	2 45	5300	200 9500	µdmAB	03.05	29N 12E	2+		04.0		5925
Nov. 12	13.22	1 23	> 7500	127 9400	µmAB	13.15	26N 04W	3+		13.7		5925
Nov. 14	02.58	2 37	8050	200 9500	µmA	02.46	27N 19W	2+		22.0		5925

Nov. 15	02.18	2 42	24000	200 9400	µdmAB	02.07	26N 33W	3		02.6	5925
Nov. 20	20.23	0 52	-	108 2800	µmA	19.55	25N 90W	2+		20.9	5925
Dec. 05	18.28	0 17	-	108 2800	µmA	18.25	27N 70E	3+		(06.)05.0	5959
1961											
Mar. 26	10.20	1 19	530	21 9400	µdmA	10.09	16S 72E	2+			6069
Apr. 04	22.34	0 11	-	200 3750	µmA	22.33	14N 17E	1			6077
June 11	10.10	> 1 30	40	30 9400	µdmAB	{ < 09.25	02N 47W	2		19.5	6135
						{ < 09.41	04N 61E	1			6140
June 11	15.03	0 27	-	18 9400	µmA	15.02	02N 50W	2+		19.5	6135
July 11	16.52	1 33	1380	108 9400	µmAB	16.15	07S 31E	2+			6171
July 12	10.18	1 42	6000	23 9400	µmA	09.50	07S 22E	3		11.3	6171
July 15	14.30	2 00	260	18 9400	µmAB	14.33	13N 15E	3		15.8	6172
						15.08	07S 21W	2		15.8	6171
July 18	09.39	0 56	> 2400	23 9400	µmAB	09.20	08S 59W	3+	09.8		6171
July 20	15.52	0 38	> 4000	108 9100	µmA	15.24	06S 90W	2+		16.3	6171
July 24	04.30	1 10	30	200 9400	µmAB	< 04.03	15N 19E	3		09.0	6178
July 28	02.27	0 23	220	200 9400	µmA	< 02.44	13N 39W	2+			6178
Sept.08	15.40	0 30	50	108 9400	µmA	14.45	19N 52W	2			6212
						15.45	10S 90E	1			6223
Sept.10	19.30	0 45	-	18 2800	µmA	19.50	16N 90W	1		20.8	6212
Sept.16	10.55	1 10	635	23 9400	µdmA	10.57	18N 77E	2+			6227
Sept.28	22.11	0 39	1600	18 9500	µmAB	22.02	13N 29E	3			6235
Nov. 10	14.30	0 30	140	18 9400	µmA	14.34	19N 90W	1+		15.0	6264a
1962											
Feb. 03	12.05	2 25	25	108 9400	µmAB	< 12.20	09N 67W	1+		20.5	6326
Mar. 01	16.35	0 10	-	108 2800	µmAB	16.34	13S 56W	2			6351
Mar. 13	14.48	> 1 10	860	108 9400	µmA	14.44	10N 66E	2+			6366
Apr. 12	21.48	0 37	75	108 9500	µmA	21.49	11N 19E	1			6386
Apr. 18	18.00	1 40	-	222 2800	µmA	17.34	10N 05E	3			6393
Apr. 22	15.32	1 03	50	19 9100	µmA	14.30	08N 48W	2+			6393
May 01	19.15	0 13	-	18 2800	µmA	19.16	19N 60E	1			6411
May 31	11.20	1 00	33	111 9400	µmA	10.46	14N 68W	2			6426
Sept.07	15.10	1 00	26	108 9400	µdmA	15.07	14S 24E	2			6548

Date	Type IV Burst Start. UT	Dur. h min	3 cm flux s.u.	Frequ. ext. MHz	Spectr.	Flare Start. UT	Position	Imp.	Particles GLE	PCA Start. UT	SSE Start. UT	Plage Reg. No.
1962												
Sept.13	09.36	1 24	10	111 9400	µmA	< 09.19	23N 51E	1+				6553
						< 09.50	15S 52W	1				6548
Oct. 23	16.42	0 14	-	221 2800	µmA	16.42	03N 70W	2		17.5		6581
1963												
Jan. 30	13.27	1 28	-	111 2800	µmA	13.05	10N 54W	1				6680
May 01	05.25	>1 00	1500	200 9400	µdmA	05.25	15N 46E	2				6790
May 20	22.54	0 31	-	108 2800	µmAB	23.03	05N 19W	1-				6805
Aug. 06	08.58	2 32	9	23 9400	µdmAB	08.55	13N 12W	2		10.3		6909
Aug. 09	22.35	0 25	140	16 9400	µmA	22.34	07N 80W	1		22.3		6908
Aug. 17	16.09	0 30	340	18 9400	µdmA	15.40	12S 09E	1+				6924
Sept.15	00.15	2 10	17000	108 9500	µdmA	00.15	15N 75E	2		14.0		6964
Sept.16	13.02	3 08	500	108 9400	µdmA	13.00	11N 49E	2		19.3		6964
Sept.20	07.14	1 36	270	536 9400	µdmA	07.13	14N 04E	2		10.2		6964
Sept.20	23.47	2 08	3850	18 9500	µmAB	23.14	10N 09W	2		24.0		6964
Sept.26	07.05	1 55	4350	23 9500	µdmA	06.38	13N 78W	3		07.5		6964
Oct. 22	13.33	0 47	260	23 9400	µmA	13.21	11N 50E	2				7003
Oct. 28	01.40	1 30	145	200 9500	µdmA	01.35	12N 24E	3		08.0		7003
1964												
Mar. 16	15.47	> 0 60		108 2800	µdmAB	15.55E	N05 W75	1		x		7182
1965												
Feb. 05	17.53	1 50		18 2800	µdmAB	17.50	N07 W25	2		(6.)00.20	18.10	7661
June 05	18.07	1 03		18 2800	µdmAB	18.07	S12 W50	1-				7842
Oct. 02	04.13	0 10	865	200 9400	µdmA	04.01	N24 E13	2+				8005
Oct. 04	09.38	>3 20	13	23 9400	µdmAB	09.35	S23 W31	2		x		8012
Dec. 29	11.16	>2 40	20	23 9400	µdmB	11.22	N11 W62	1		x		8105
Dec. 30	00.19	0 35	40	200 9400	µdmA	00.06	N09 W71	1+			14.30	8105

1966											
Jan. 17	10.10	>4 00	354	23 9400	μdmAB	10.32E	N19 E27	2B			8131
Mar. 19	03.39	0 20	3640	500 9400	μmA	03.27	N20 E33	3N			8207
Mar. 20	09.42	4 00	2590	18 9400	μmAB	09.28	N22 E15	3B		x	8207
Mar. 24	02.30	0 14	930	18 9400	μmA	02.25	N18 W17	3B	x		8207
Mar. 25	10.05	0 23	180	23 9400	μmA	10.00	N15 W57	2B			8207
Mar. 30	12.47	>3 20	2630	23 10700	μmA	12.41	N27 W47	2N			8223
July 07	00.18	1 54	12750	200 17000	μdmAB	00.22E	N34 W45	2B		00.58	8362
July 09	02.29	2 50	385	200 9400	μdmAB	02.30	N35 W75	1B			8362
Aug. 28	11.52	>3 30	18	23 9400	μmB	11.53	N08 W60	SN			8454
Aug. 28	15.22	>1 40	3580	18 9400	μdmAB	15.22	N23 E04	3B		(29.)07.00	8459
Aug. 30	14.54	1 59	835	18 10700	μdmAB	14.50	N22 W27	1B			8461
Aug. 31	00.36	0 43	235	500 9400	μ(mA)	00.36	N22 W29	1B			8461
Sept.02	05.36	1 55	6970	111 17000	μdmAB	05.41	N22 W57	2B		(14.00)	8461
Sept.03	10.03	3 07	230	64 9400	μdmA	10.14E	N15 W75	1N			8461
Sept.04	04.13	>0 15	2640	200 17000	μdmA	04.10	N22 W84	3B			8461
Sept.12	09.25	>6 10	30	23 9500	μdmB	09.25	N13 E90	2N			8505
Sept.14	10.12	0 38	14	23 9400	μdmAB	10.12	S19 W90	1B		x	8484
Sept.18	14.51	1 10	840	208 10700	μdmA	14.52	N23 E42	1B			8509
Sept.20	10.47	>5 30	400	23 15000	μmA	10.46	N10 W14	3B		x	8505
Sept.21	09.30	>4 30	450	1500 15000	μ	09.29	N21 E03	2N			8509
Oct. 14	05.20	0 21	625	612 17000	μ	05.24	N18 E47	-B			8546
1967											
Jan. 12	02.27	55	1070	500 17000	μ(mA)	02.31	N16 W90	1	13.50		8631
Feb. 01	12.15	1 32	320	23 9400	μdmAB	(13.32	N13 W90	1			8659)
Feb. 01	23.04	0 36	285	500 3750	μ(mA)	23.20	N28 E60	1-	(2.)20.00		8680
Feb. 03	02.45	2 35	200	(10) 9400	μdmAB	02.41	S28 W53	1-			8667
Feb. 04	16.38	3 10	120	(7.6) 10700	μmB	16.40	N11 E40	2B			8682
Feb. 05	12.07	>3 00	16	23 9400	μdmB	12.00-14.05 no flare patrol					
Feb. 27	16.35	1 25	6200	200 10700	μdmA	16.36	N21 E22	1-		03.00	8707
Mar. 04	17.16	0 20	1330	606 10700	μd(mA)	17.16	N24 W68	1+			8704
Mar. 20	11.50	1 12	320	600 9500	μ	11.55E	S14 W23	-F			8724

Date	Type IV Burst Start. UT	Dur. h min	3 cm flux s.u.	Frequ. ext. MHz	Spectr.	Flare Start. UT	Position	Imp.	Particles GLE	PCA Start. UT	SSB Start. UT	Plage Reg. No.
1967												
Mar. 22	00.23	>2 20	2000	408 17000	µd(mA)	00.22	N25 E70	3B				8740
Mar. 23	19.19	0 55	130	(10) 10700	µdmB	19.14	N25 E32	1B				8740
Apr. 25	09.22	0 55	90	111 9400	µmA	09.27	N35 W70	-F				8776
May 03	15.38 (+16.06)	>8 00	26	(10) 9400	µdmB	15.37	N25 E51	2B				8798
May 18	18.30	>0 10	40	(10) 10700	µmB	18.54	S18 W18	-N				8809
May 18	23.50	0 50	140	208 9400	µmA	23.37	N23 E72	-N				8818
May 19	15.24	1 10	310	(10) 9400	µdmA	15.24	N24 E64	1B				8818
May 20	15.10	1 30	555	(10) 9139	µdmB	15.09	N22 E51	2B				8818
May 21	19.22	>1 00	1970	(10) 10700	µdmB	19.21	N24 E39	2B				8818
May 23	18.35	>4 25	23000	(10) 8800	µdmAB	18.35	N28 E24	3B		24.00	19.25	8818
May 24	02.55	4 40	40	30 9400	µdmB	02.57	N26 E89	-F			x	8824
May 24	08.03	1 21	50	200 9500	µd(mA)	08.00E	S27 E77	-N				8825
May 25	10.38	>3 20	50	23 9100	µdmB	10.39	N24 W07	2N				8818
May 26	12.26 (+15.30)	>4 00	290	30 10700	µdmB	12.28	N30 W04	2B				8818
May 28	05.28	1 40	7300	23 17000	µdmAB	05.27	N28 W33	3B		06.05	06.25	8818
June 03	02.25	>1 25	85	500 9400	µd	02.26	N23 E13	1B			x	8831
June 23	18.55	~1 40	27	(10) 10700	µdmA	18.54	N19 E23	1N				8863
July 25	14.27	>1 00	1170	600 10700	µ(mA)	14.24	N28 E39	1B				8905
Aug. 26	00.13	1 35	21	(10) 9400	µdmAB	00.13	S19 E00	1B				8949
Sept. 18	23.15	>4 00	28	(10) 9400	µdmAB	23.16	N16 W60	2B		x	x	8973
Oct. 29	02.56	0 40	210	720 17000	µd(mA)	02.58	N09 W80	1N				9034
Oct. 29	23.42	1 00	4290	200 17000	µmAB	23.47	N10 W90	1B			x	9034
Nov. 02	08.54	~0 20	~500	30 15000	µdmA	08.55	S18 W02	1B		x	12.00	9047
Nov. 04	11.52	0 15	36	23 9100	µdmA	11.52	S18 W33	1B		x	14.00	9047
Nov. 10	17.49	1 50	13	(10) 8800	µmB	17.52	N16 E42	-N			21.23	9065
Dec. 11	23.46	0 35	170	200 9400	µdmA	23.47	S21 W18	2B				9108

Dec. 13	13.40	1 15	24	(23)	10700	µmB	13.39	S22 W36	2B			9108
Dec. 16	02.46	1 30	31	208	9400	µd(mAB)	02.47	N23 E66	2N	x	05.36	9118
1968												
Jan. 02	05.21	0 14	3970	200	9400	µdmA	05.19	S22 E89	1B			9145
Feb. 10	16.04	0 40	76	(10)	8800	µdmAB	17.56	N20 E46	-F		(11.)22.16	9204
June 03	23.32	0 50	16	(10)	9400	µdmE	23.31	S30 W15	1N			9420
June 09	08.39	1 20	1360	23	71000	µdmAB	08.35	S14 W08	3B	11.00	10.00	9429
June 11	09.34	0 11	540	111	19000	µdmA	09.34	S14 E34	1B			9449
July 06	09.40	1 45	6900	(23)	71000	µdmA	09.46	N14 E89	1N		x	9503
July 08	17.05	0 50	5100	18	19000	µdmAB	17.06	N13 E53	1B	x		9503
July 09	18.06	0 30	520	600	19000	µ(mA)	18.08	N13 E40	2B	x	x	9503
July 12	13.43	>1 40	120	18	15400	µdmAB	13.41	N11 W21	2N	x	x	9499
Aug. 14	13.26	1 20	465	23	22000	µdmA	13.26	N13 W80	1B		x	9567
Sept.04	00.30	1 13	30	184	9400	µdmB	00.28	N13 W14	1N			9630
Sept.26	00.25	40	920	200	9400	µdmA	00.26	N14 E35	2B	x	x	9687
Sept.28	01.47	50	180	208	9400	µdmA	01.50	N13 E06	2B	x	x	9687
Sept.29	09.30	1 40	205	23	19000	µdmA	09.22	N13 W13	2B	}		9687
Sept.29	16.19	0 30	2810	18	19000	µdmA	16.18	N17 W52	2B	} 20.00	16.50	9678
Sept.30	06.45	>1 40	10	23	9400	µdmAB	06.51	N12 W11	1N			9687
Oct. 02	20.12	~0 50	145	18	15400	µmA	20.14	N19 E68	1B			9705
Oct. 03	23.43	3 30	85	184	9400	µdmAB	23.48	S16 W37	2B	x	x	9692
Oct. 04	21.23	1 10		(18)	2800	µdmB	21.20	N20 E40	2B			9705
Oct. 23	23.54	1 00	> 225	208	9400	µmA	23.56	S12 E59	2B			9740
Oct. 24	20.45	1 35		184	2800	µmB	20.53	S18 E52	1B			9740
Oct. 26	01.40	1 30		612	3750	$(\mu)d^+$	01.09	S20 E32	1N			9740
Oct. 27	12.31 (+13.06)	> 2 10	1680	30	15400	µmB	12.35	S17 E16	1B			9740
Oct. 29	11.32 (+12.14)	>3 10	780	23	19000	µdmB	{11.14 {12.16	S16 W12 S16 W10	1B		x	9740 9740
Oct. 29	15.16	1 20	670	23	19000	$\mu d^+ mB$	15.15	S14 W19	-N		(30.)05.00	9740
Oct. 30	12.30	>2 20	24	23	15400	µdmAB	12.34	S18 W25	2B			9740
Oct. 30	23.40	1 00	2000	200	17000	µdmAB	23.39	S14 W37	3B	(31.)13.30	(31.)10.00	9740
Oct. 31	22.31	1 40	2270	184	17000	µdmA	22.32	S15 W49	2N	(01.)01.00	x	9740

	Type IV Burst					Flare			Particles			Plage
Date	Start. UT	Dur. h min	3 cm flux s.u.	Frequ. ext. MHz	Spectr.	Start.	Position	Imp.	GLE	PCA Start. UT	SSE Start. UT	Reg. No.
1968												
Nov. 01	08.15	3 20	1930	30 19000	µdmAB	08.14	S18 W46	2B			x	9740
Nov. 01	19.45	0 45	3390	184 15400	µdmA	20.02	S13 W57	1B			x	9740
Nov. 02	09.24	2 05	6600	23 19000	µdmAB	09.26	S13 W64	-N		x	x	9740
Nov. 04	05.12	~1 25	8000	612 17000	µdmA	05.15	S15 W90	2B		x	x	9740
Nov. 11	14.38	0 34	12	(10) 15400	µdmA	13.13	N18 E04	-N				9760
Nov. 18	10.26	>3 30	2970	23 19000	µdmAB	10.26	N20 W90	1B	10.45	10.40	x	9760
Nov. 22	00.50	1 00	17	80 9400	µdmAB	00.56	S16 E40	1N				9780
Dec. 02	21.00	0 30		(10) 2700	µmA	21.15	N21 E89	1N				9802
Dec. 11	11.44	1 33	420	111 19000	µdmA	11.46	S20 W49	-F				9801
Dec. 24	22.27	0 10		(10) 2700	µdmA	22.28	N19 E29	1B				9842
Dec. 27	10.48	>1 40	3090	30 71000	µdmAB	10.46	N16 E02	2B				9842
Dec. 29	17.50 19.15	4 00	425	184 15400	µdmAB	19.14	N16 W33	1B				9842
1969												
Jan. 17	17.00	0 20	1360	18 15400	µdmA	17.04	N17 E46	-B				9873
Feb. 20	05.58	>1 00	900	650 17000	µd(mA)	06.19	N15 E38	2B				9946
Feb. 23	04.42	1 00	785	200 9400	µd(mA)	04.41	N12 W09	1B				9946
Feb. 24	22.46	1 12	975	(10) 9400	µdmA	22.50	N12 W31	2B				9946
Feb. 25	08.55	2 45	6200	30 9400	µdmB	09.03	N13 W37	2B		09.11	09.40	9946
Feb. 25	16.45	>0 20	690	(10) 10700	µdmA	16.53	N12 W40	1N			x	9946
Feb. 25	19.32	1 15	21	184 10700	µdmA	19.37	N14 W43	1B			x	9946
Feb. 26	04.13	>0 45	3600	100 17000	µdmB	04.14 04.19	N05 E85 N13 W46	1N 2B		x	x	9963 9946
Feb. 27	14.00	>2 30	2480	(10) 10700	µdmAB	13.52	N13 W65	2B		x	x	9946
Mar. 01	19.54	0 57	155	184 10700	µmA							
Mar. 01	21.36	1 40	160	184 10700	µdmA	21.04	N08 W89	1N				9946
Mar. 12	17.24	>1 40	4000	18 35000	µdmAB	17.38	N12 W80	2B		x	x	9966
Mar. 20	11.23 14.47	>10 00	76	18 15400	µdmB	14.10	S31 E36	-F				10001

Mar. 21	01.21	3 00	1180	200	9400	µdmAB	01.41	N19 E16	2N		x	x	9994
Mar. 21	12.40	4 50	8000	(10)	35000	µdmAB	13.07	N19 E09	2B				9994
Mar. 21	18.15	3 00	950	(10)	15400	µdmAB	18.23	N20 E15	-N			x	9994
Mar. 26	12,40	5 50	325	600	35000	µ(mA)							
Mar. 27	12.55	>1 40	4050	(10)	35000	µdmAB	13.23	N21 W68	2B				9994
Mar. 30	02.47	3 20	25900	200	17000	µdmAB	03.11	N10 E53	-N	05.30	07.46	04.00	10014
Apr. 10	03.55	2 25	590	200	17000	µdmA	04.10	N11 E90	1N	}	x	x	10035
Apr. 10	10.43	1 08	90	(10)	9400			?					
Apr. 21	20.03	2 40	3900	245	35000	µdmA	19.59	N26 W64	1N				10035
Apr. 24	03.00	>4 30	195	612	17000	µ(mA)	03.13	N22 W51	2N				10035
Apr. 26	22.59	3 45	820	(10)	15400	µdmAB	22.58	N08 E38	2N				10057
May 02	17.44	1 20	520	(10)	35000	µdmAB	17.45	N08 W40	1B				10057
May 06	02.36	1 34	1300	200	9400	µdmA	02.34	S31 E67	1B				10078
May 17	19.22	0 37	1600	18	35000	µmA	19.23	N11 E90	-B				10109
May 22	18.55	1 53	330	18	35000	µdmA	19.00	N12 E40	2B				10109
May 24	07.05	>2 00	12	100	9100	µd(mA)	07.06	N12 E01	-F				10109
May 26	11.00	1 35	41	234	15400	µdmA	11.00	N11 E04	-N				10109
June 05	09.30	4 10	5800	15	71000	µdmAB	09.53	N12 E63	3B				10134
June 05	14.10	3 20	135	(10)	35000	µdmAB	14.12	N18 E75	-N				10134
June 07	09.25	1 50	200	235	15400	µdmA	07.27	N10 E34	-N				10134
June 11	16.10	2 05	265	(10)	35000	µdmA	16.16	N10 W20	2B				10134
July 03	15.08	0 55	370	184	35000	µdmA	15.15	S11 W33	1B				10166
Sept.25	14.10	2 40	160	(10)	15400	µdmA	14.12	N06 W40	1N				10325
Sept.27	03.20	4 00	90	100	9400	µdmAB	03.11	S13 E37	1F			(28.)05.00	10337
Nov. 02	09.55	4 00	720	30	19000	µmAB	09.39	N15 W90	1B		12.00	x	10385
Nov. 04	04.08	2 12	160	200	9400	µdmA	04.00	N22 E79	1B				10412
Nov. 18	16.00	4 00	6100	(10)	35000	µdmAB	16.35	N14 E40	2B				10432
Nov. 19	05.27	1 19	1260	200	9400	µdmA	05.30	N14 E33	1B				10432
Nov. 24	09.13	5 00	6370	30	71000	µdmA	09.14	N15 W32	2B			x	10432
Nov. 27	19.11	1 00	3290	245	35000	µdmA	{19.15	S13 E40	-B				10449
							19.28	N18 W83	2B				10432
Dec. 18	14.45	2 20	7	245	8800	µdmAB	-					x	

Appendix 2

Compilation of spectral diagrams of selected type IV bursts

Remarks to the figures

— The isolines are given in solar units, i. e. $\frac{W}{m^2\,Hz} \cdot 10^{-22}$.

— In general observed frequencies are indicated by horizontal straight lines, contours between these lines are due to interpolation.

— The observed frequencies are denoted in MHz. Numbers without station abbrevations refer to HHI observations. The station abbrevations are taken from the Quarterly bulletin code.

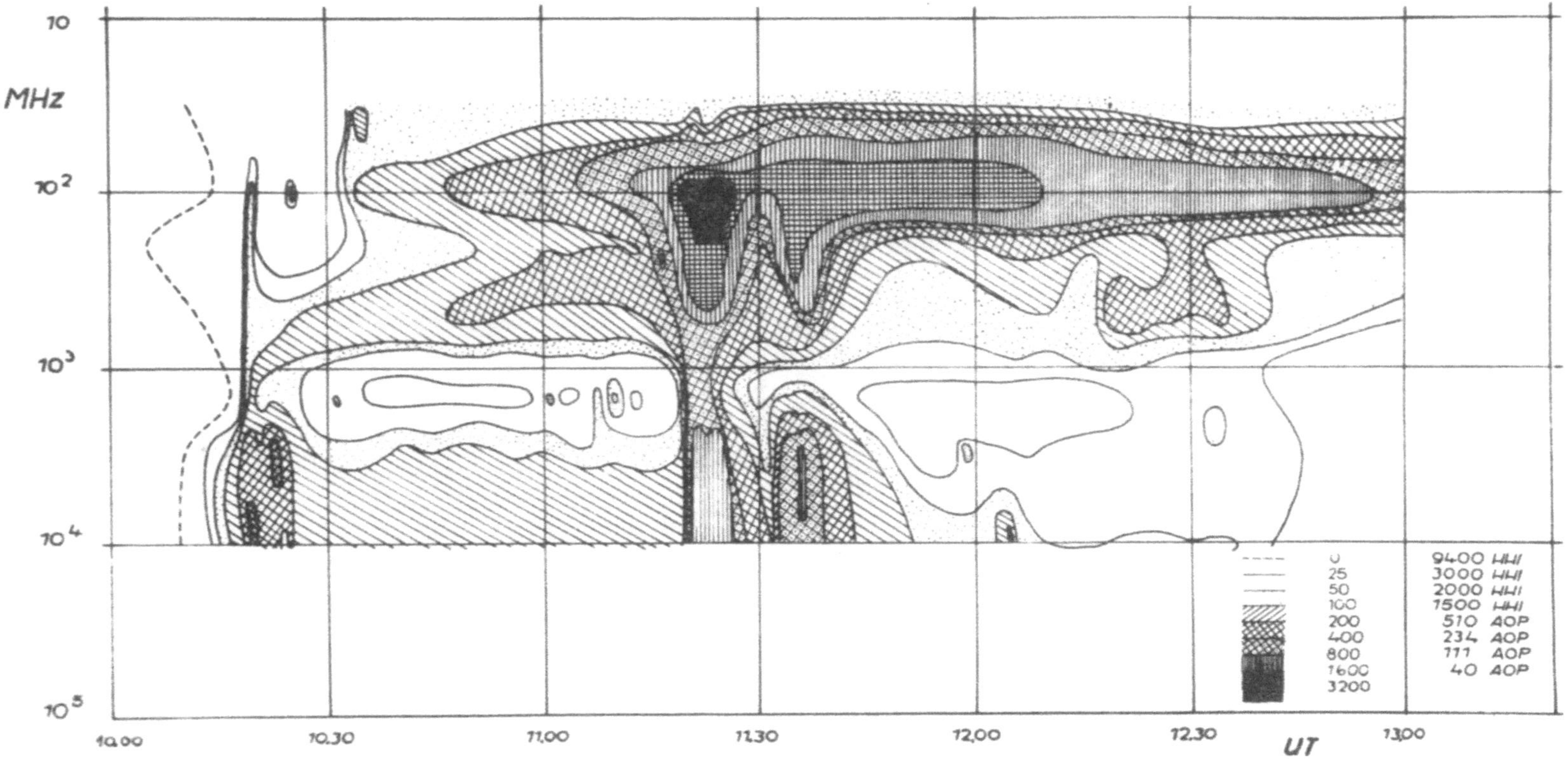

Fig. 31. Spectral diagram of burst of 1960 November 10

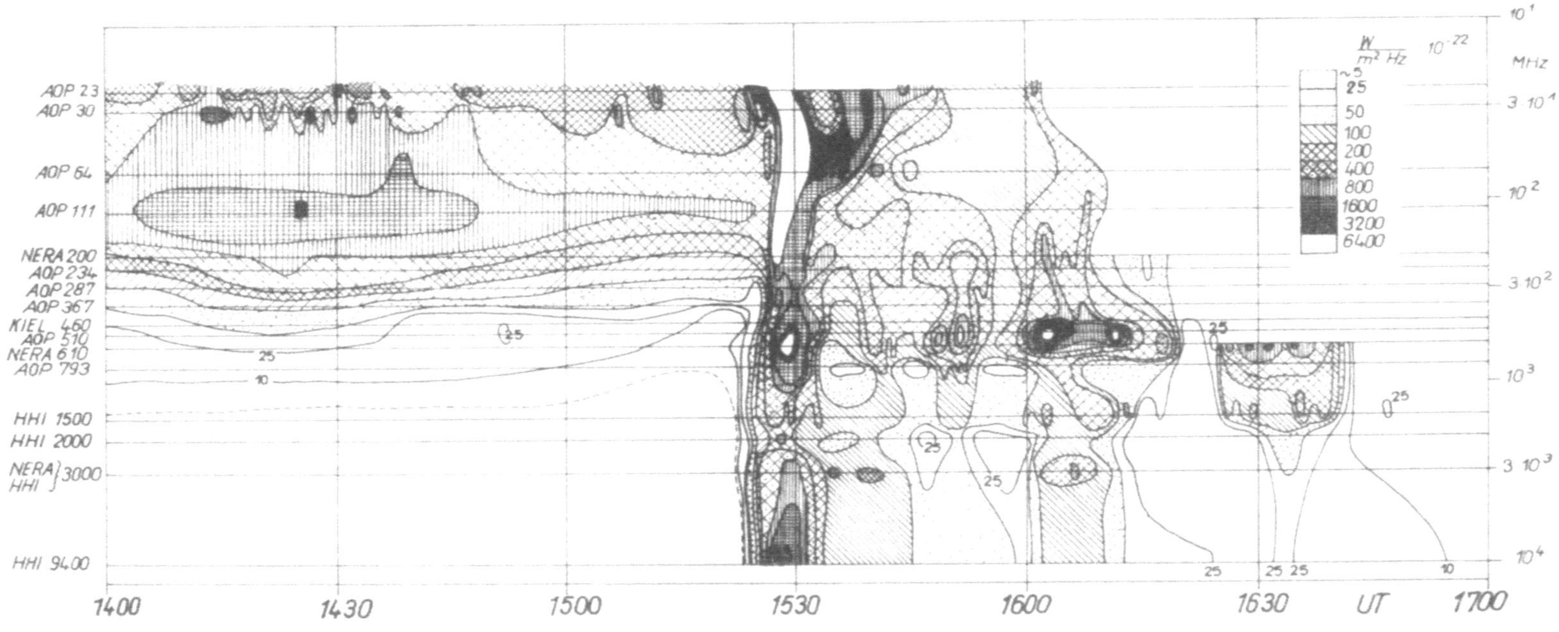

Fig. 32. Spectral diagram of burst of 1966 August 28

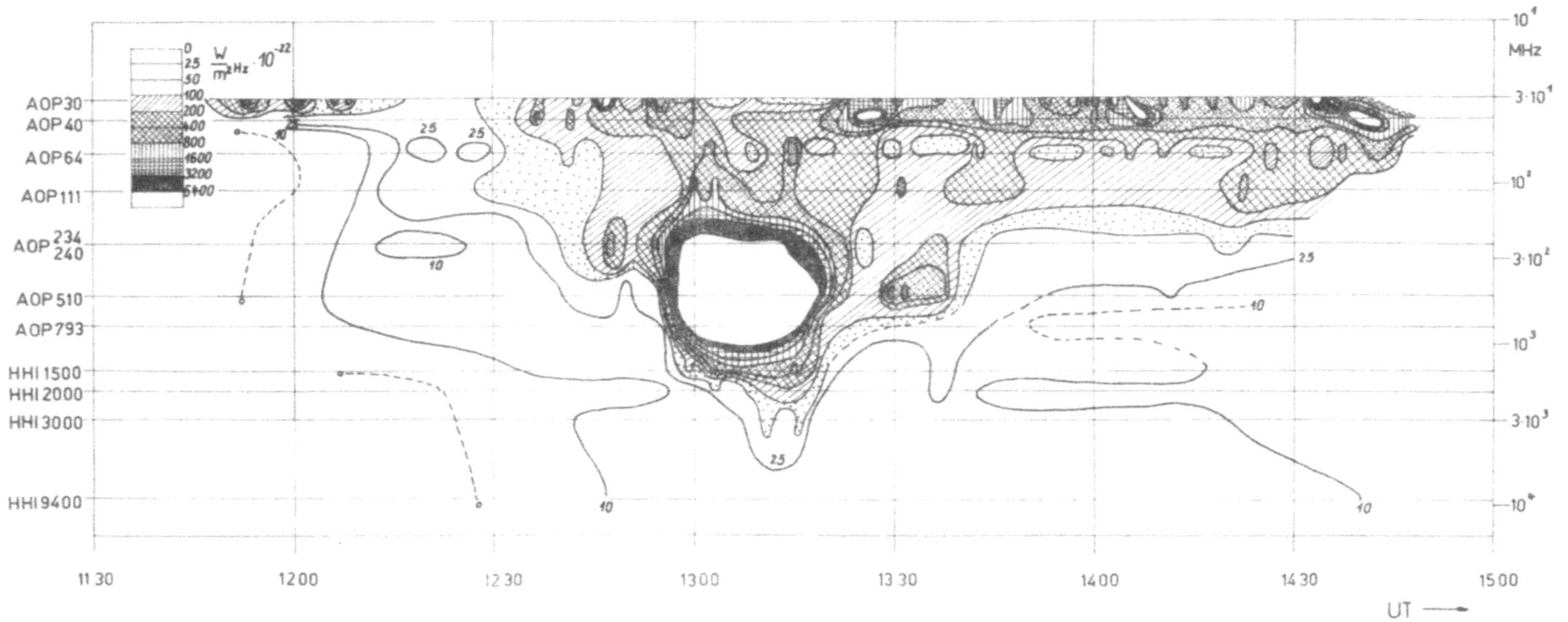

Fig. 33. Spectral diagram of burst of 1967 February 05

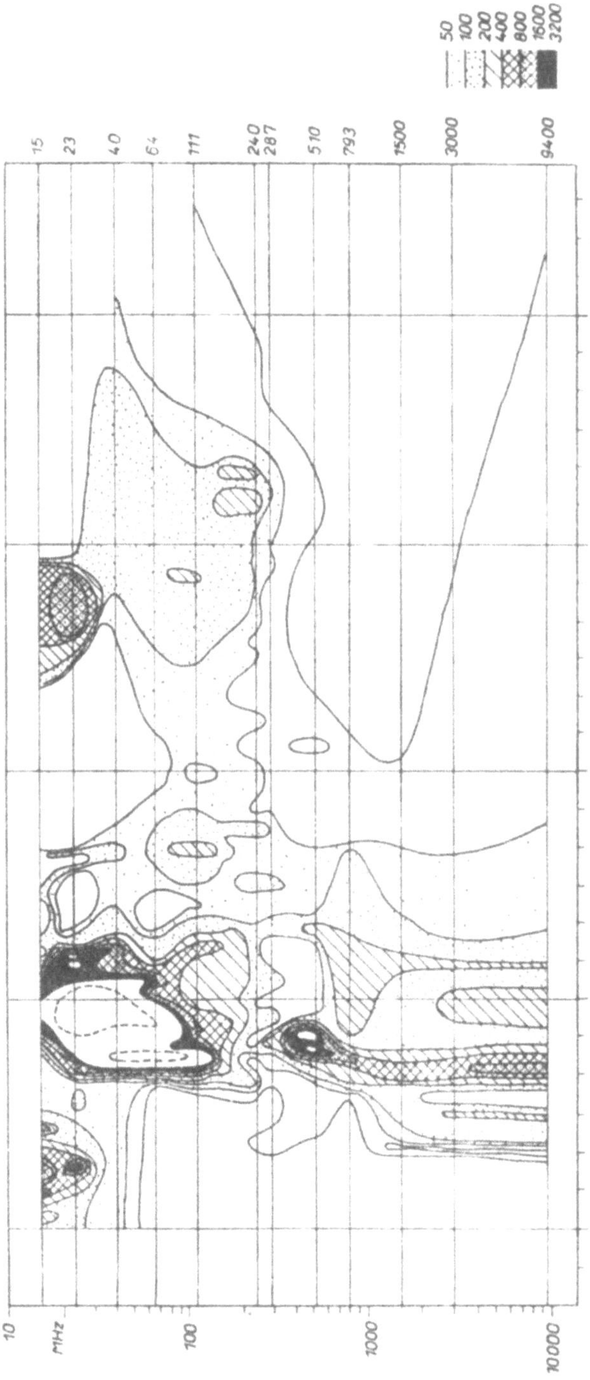

Fig. 34. Spectral diagram of burst of 1968 June 09

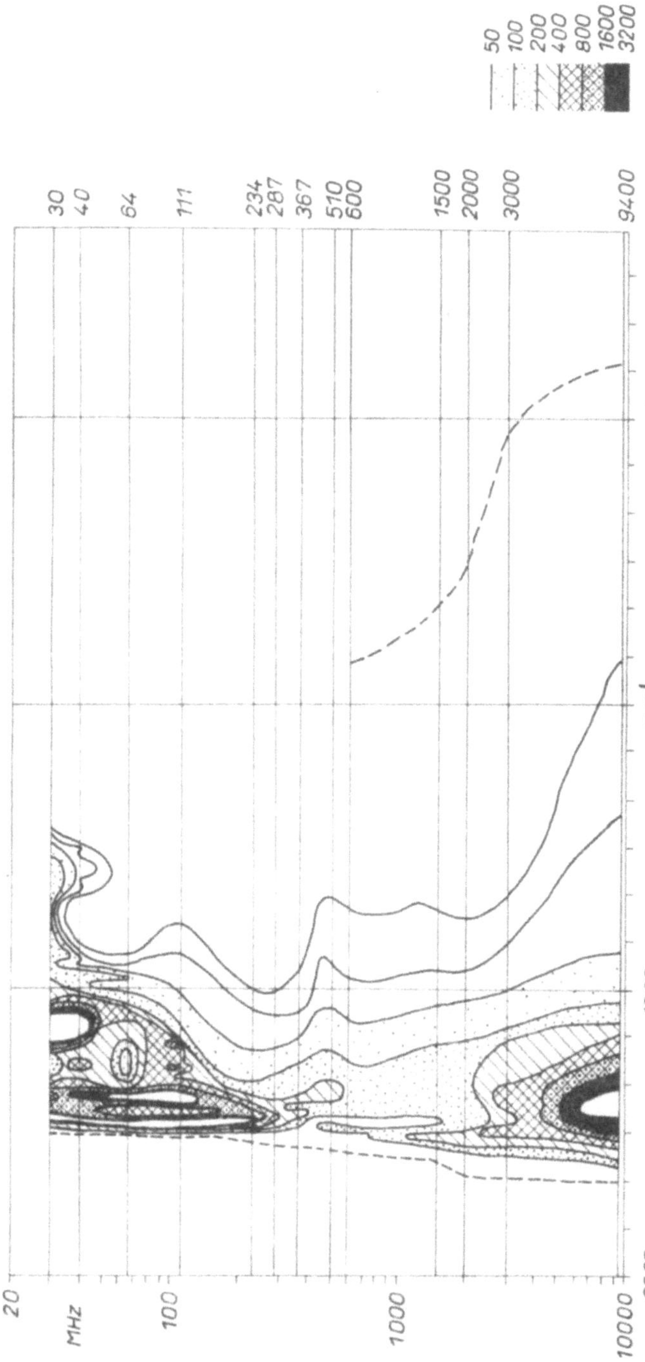

Fig. 35. Spectral diagram of burst of 1968 July 06

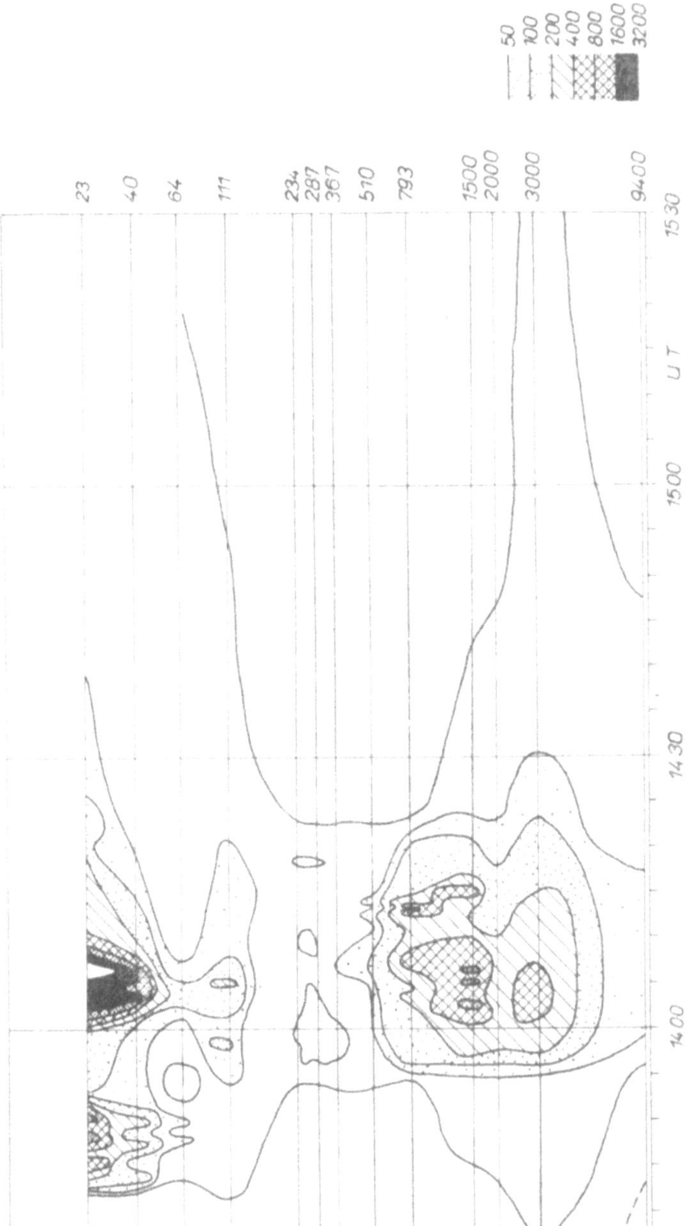

Fig. 36. Spectral diagram of burst of 1968 July 12

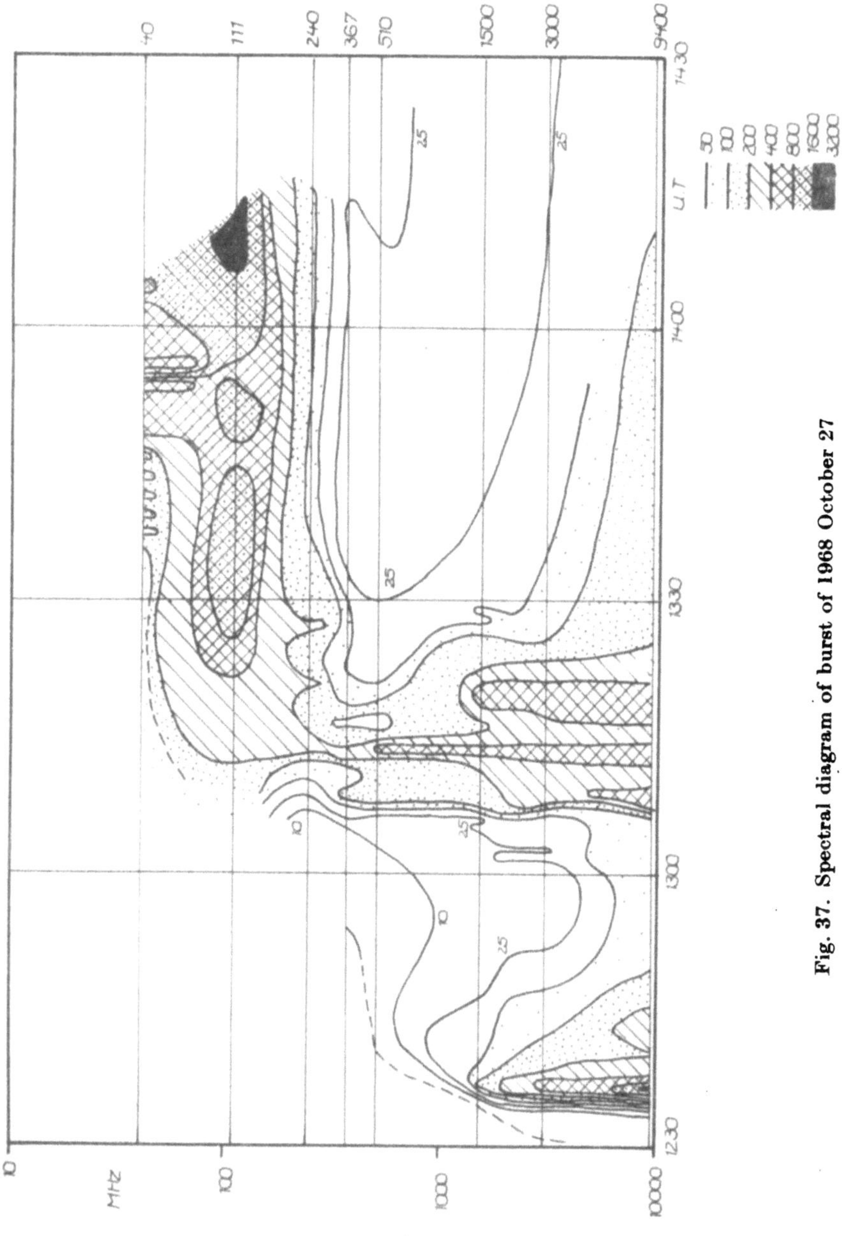

Fig. 37. Spectral diagram of burst of 1968 October 27

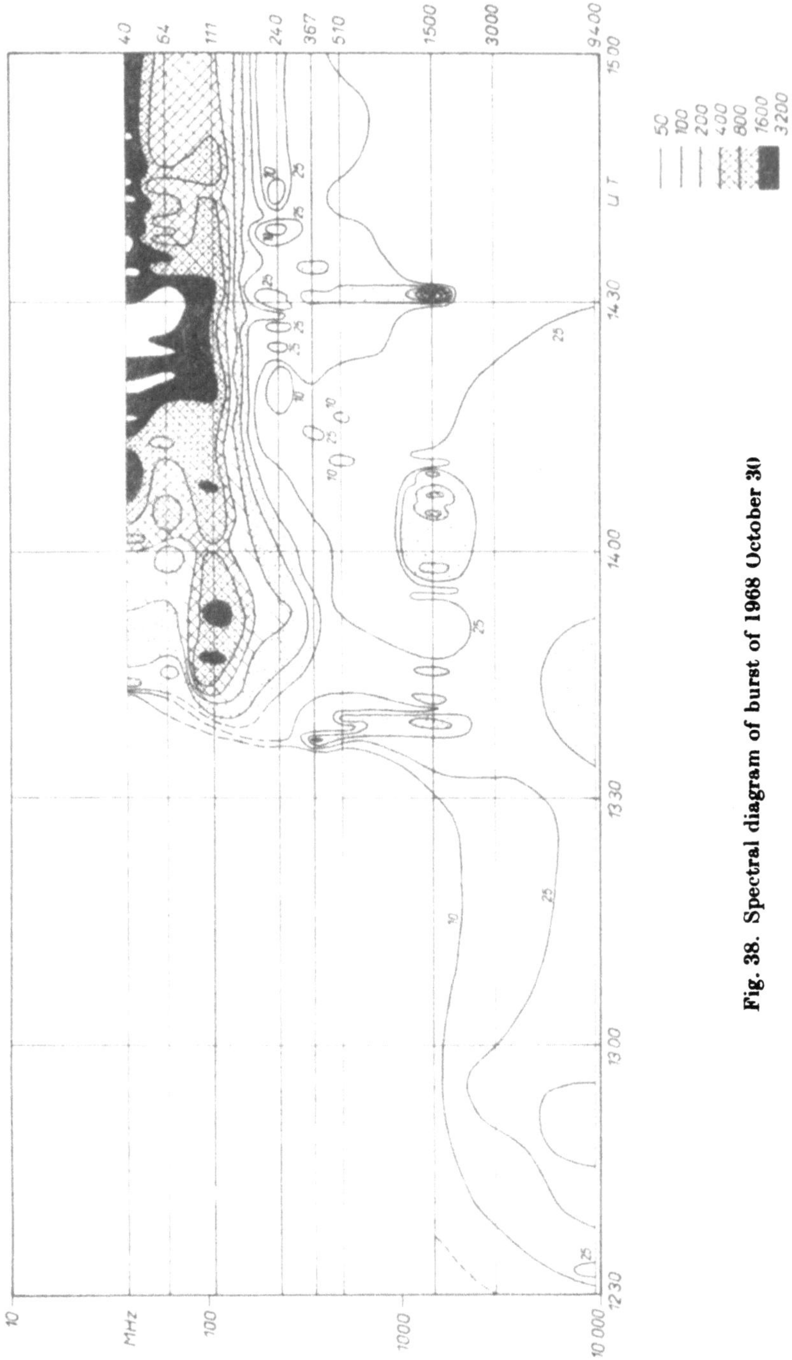

Fig. 38. Spectral diagram of burst of 1968 October 30

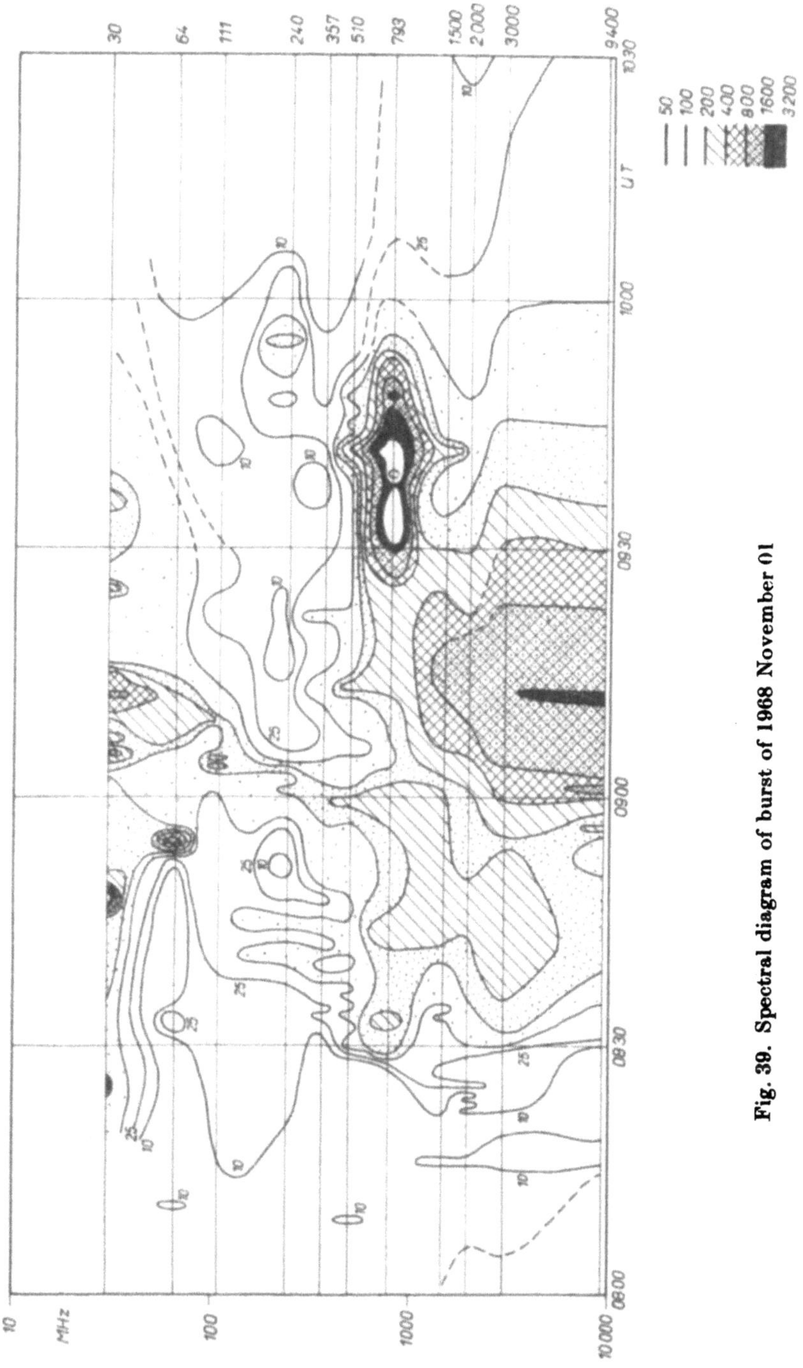

Fig. 39. Spectral diagram of burst of 1968 November 01

Fig. 40. Spectral diagram of burst of 1968 November 02

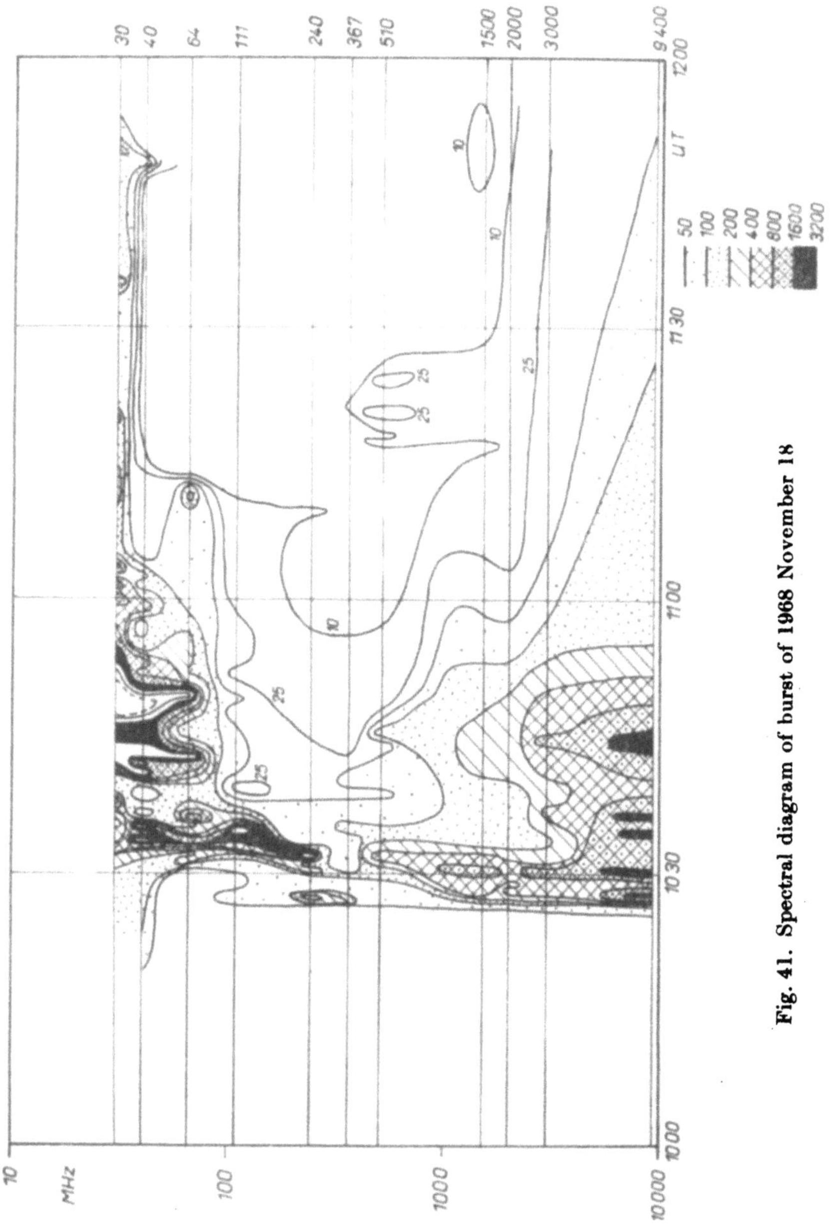

Fig. 41. Spectral diagram of burst of 1968 November 18

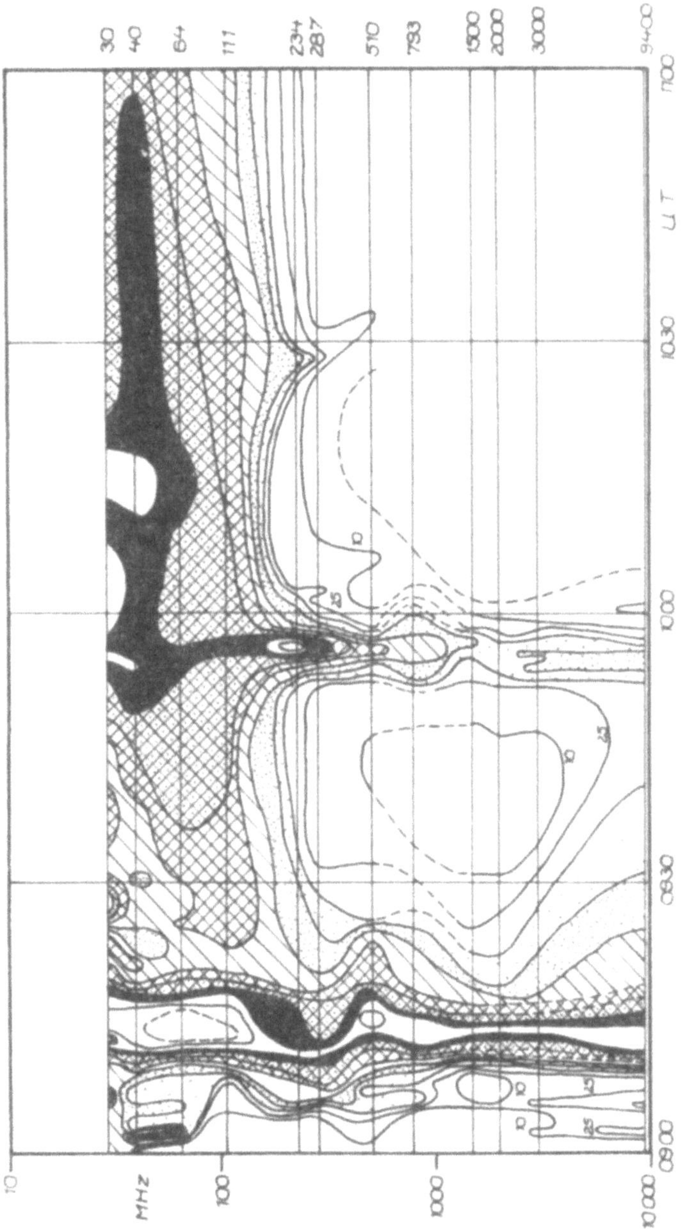

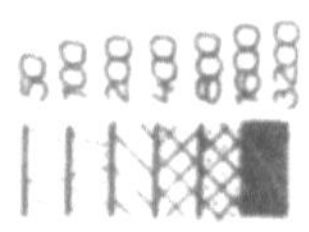

Fig. 42. Spectral diagram of burst of 1969 February 25

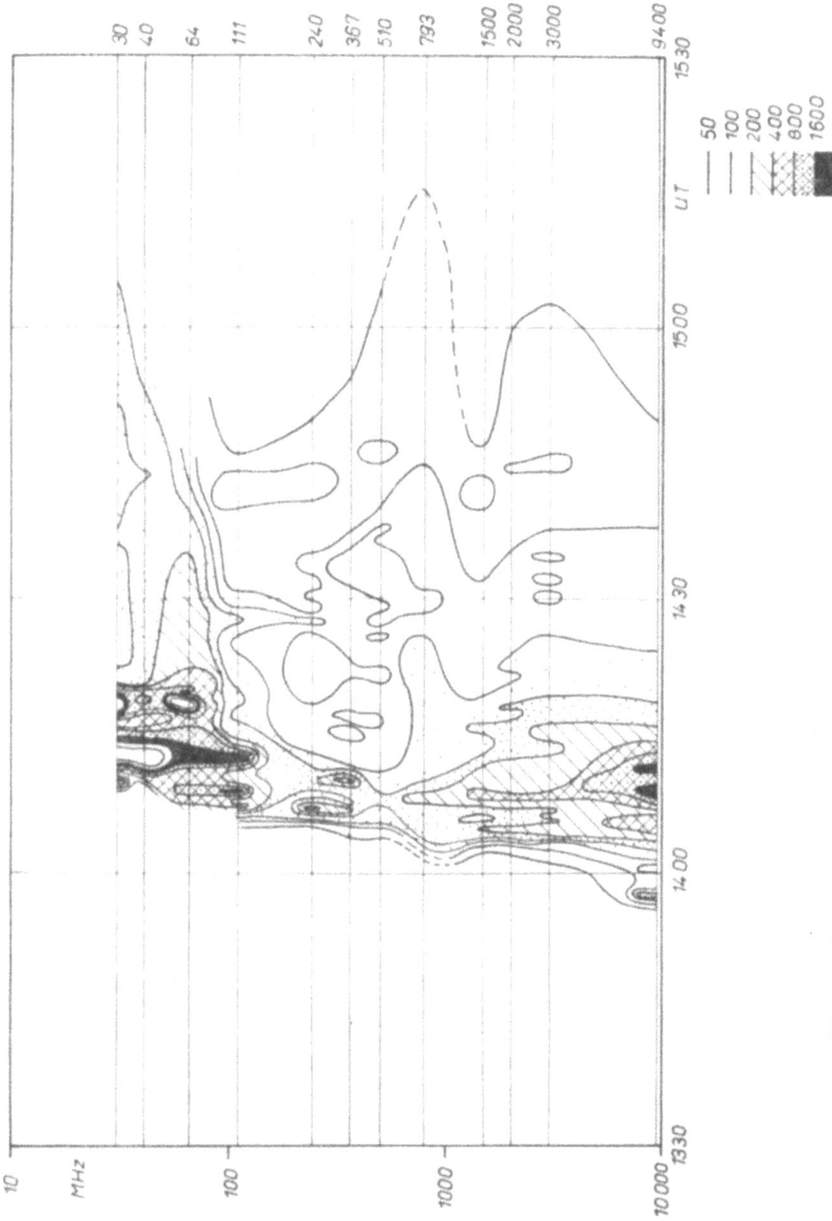

Fig. 43. Spectral diagram of burst of 1969 February 27

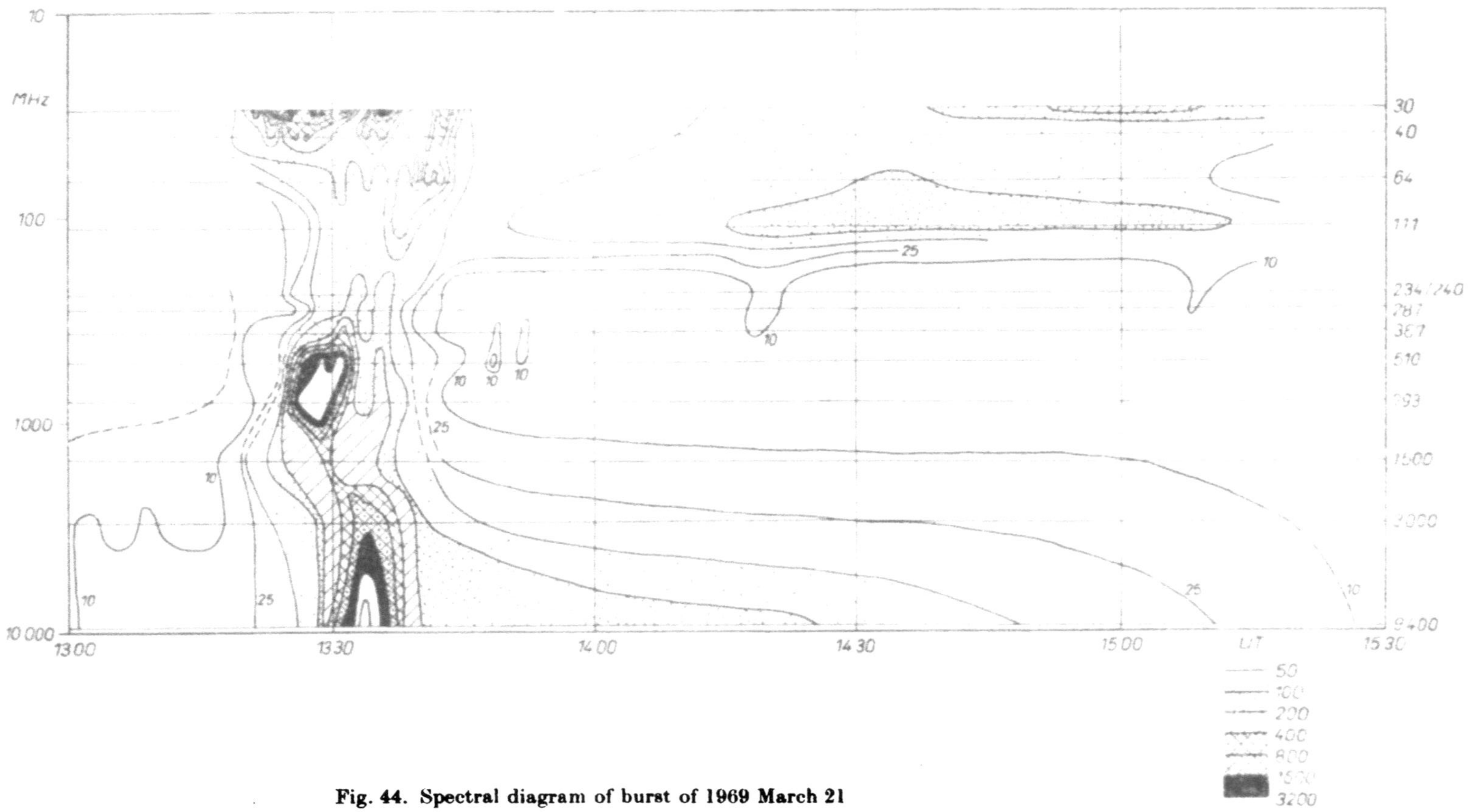

Fig. 44. Spectral diagram of burst of 1969 March 21

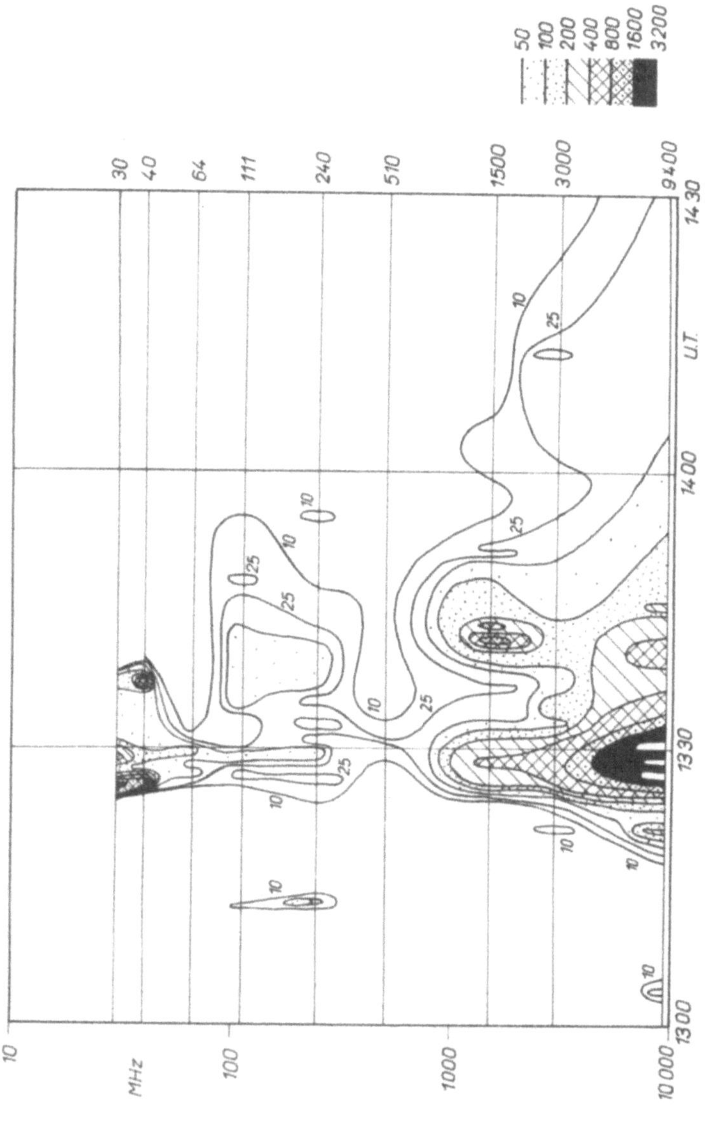

Fig. 45. Spectral diagram of burst of 1969 March 27

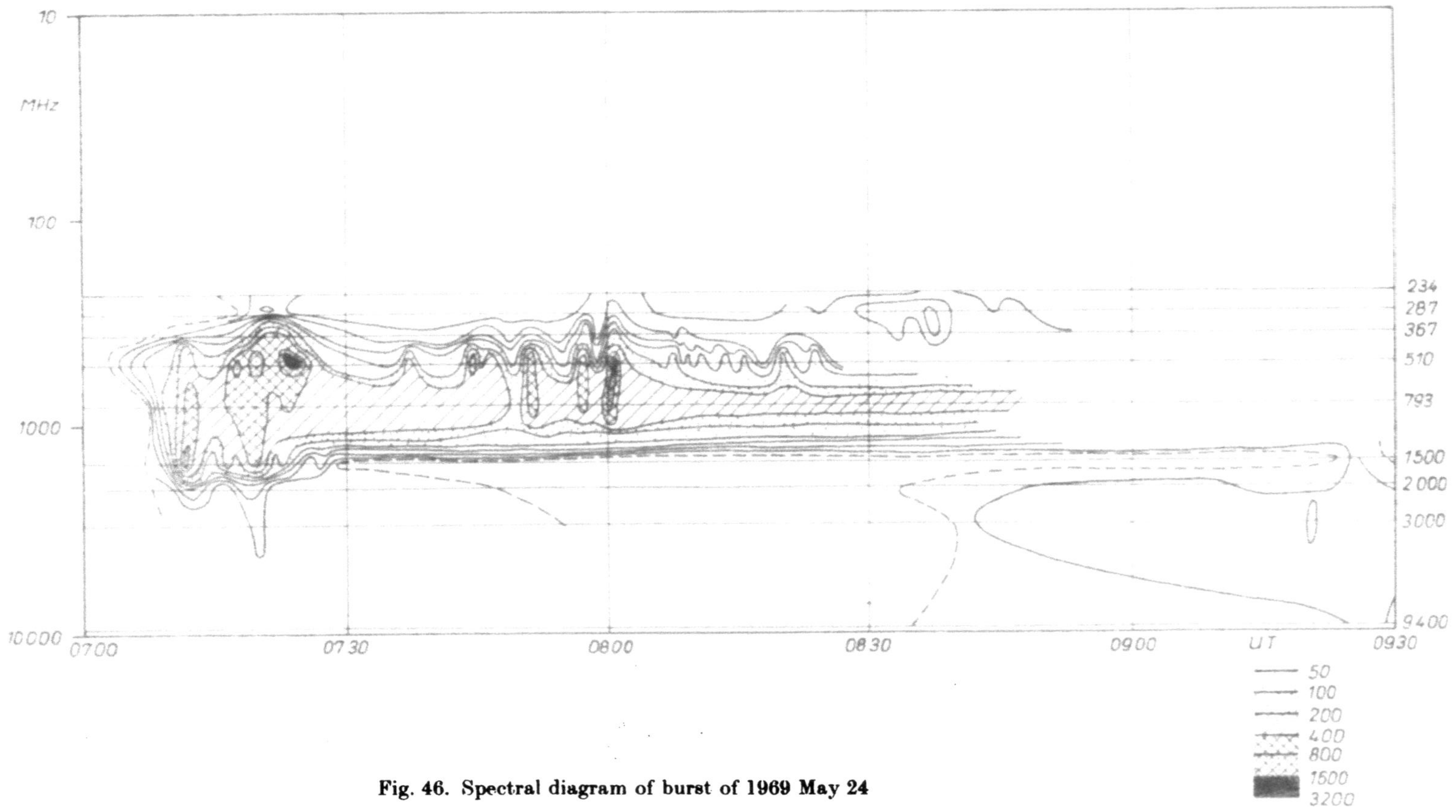

Fig. 46. Spectral diagram of burst of 1969 May 24

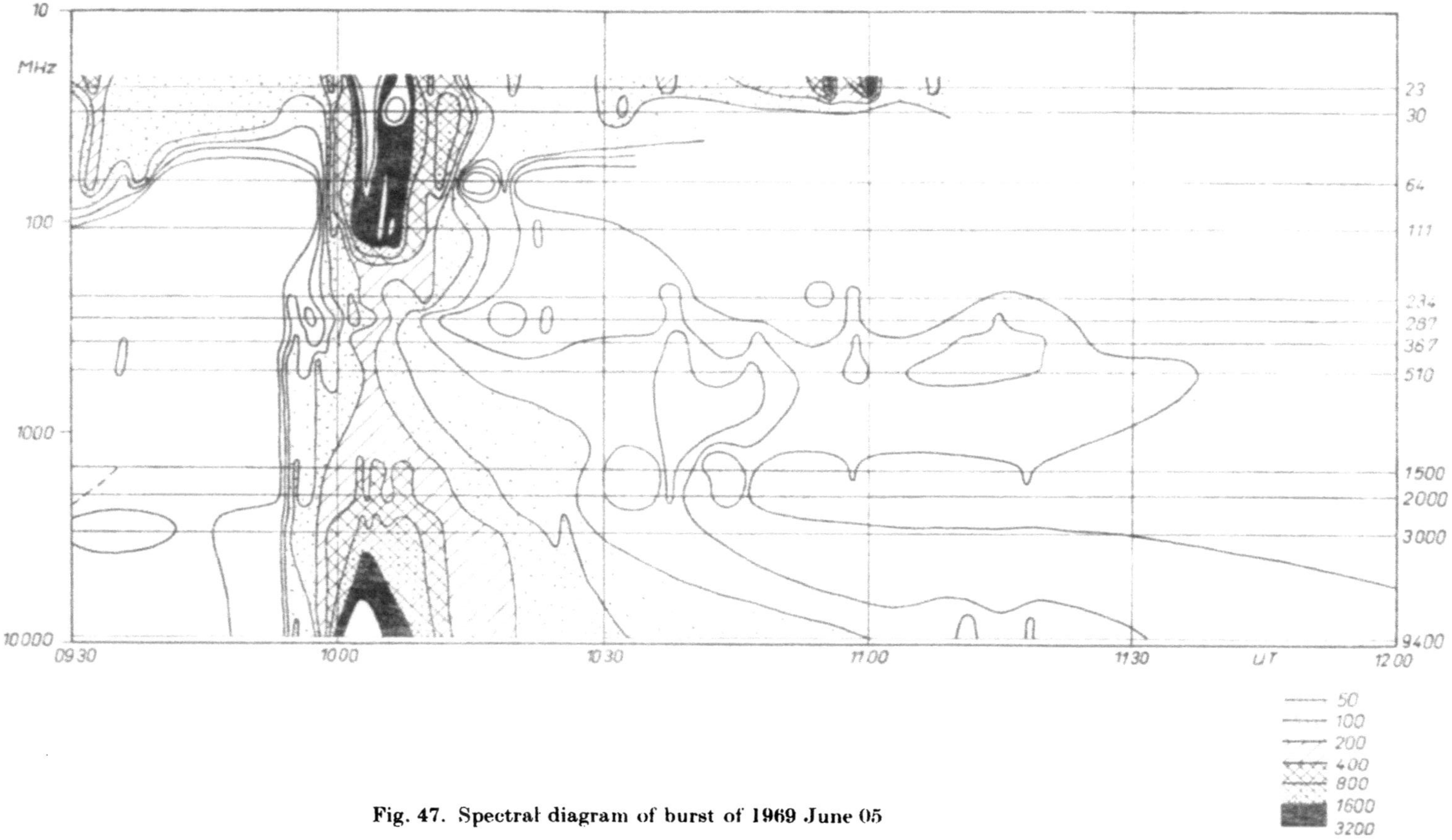

Fig. 47. Spectral diagram of burst of 1969 June 05

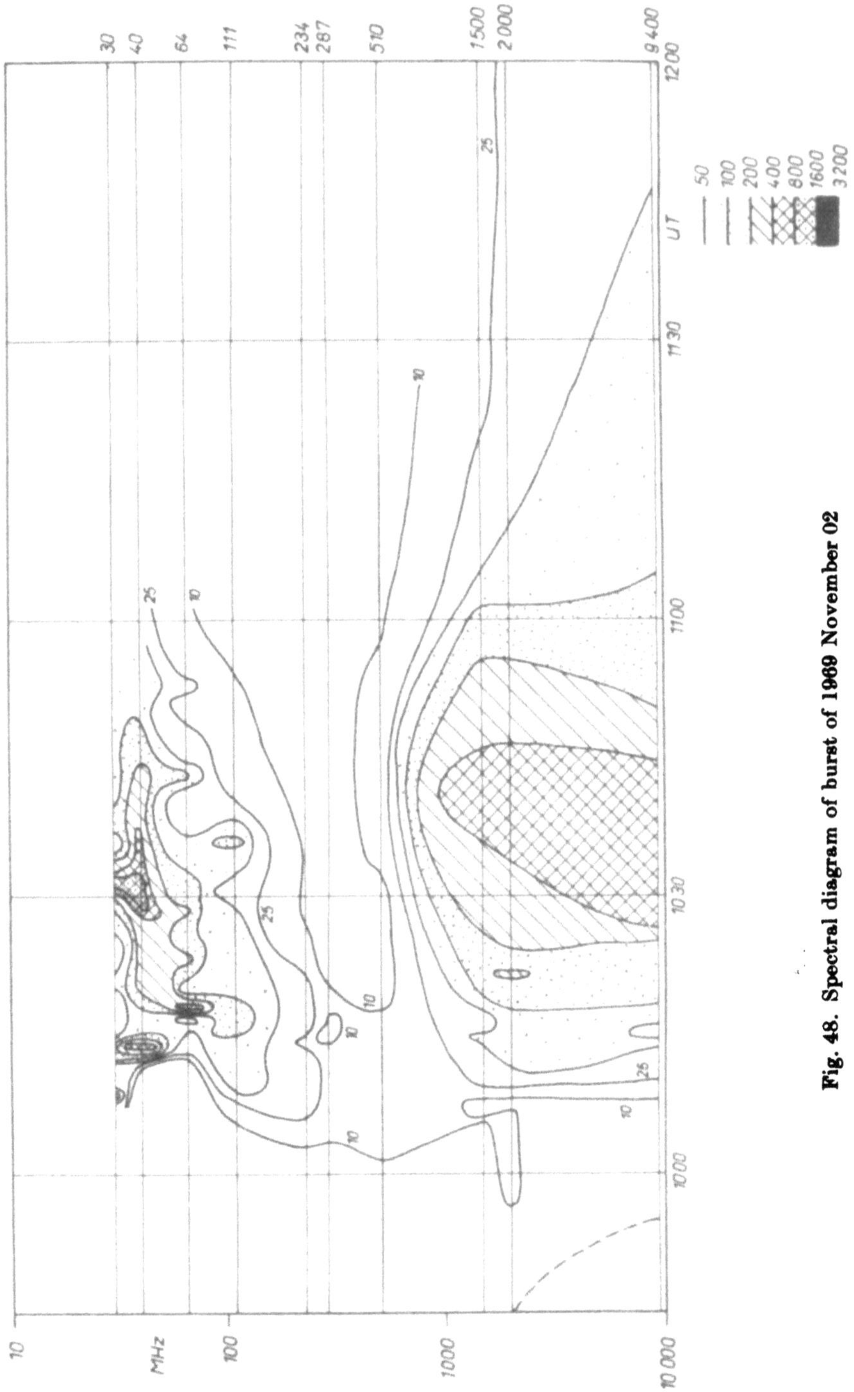

Fig. 48. Spectral diagram of burst of 1969 November 02

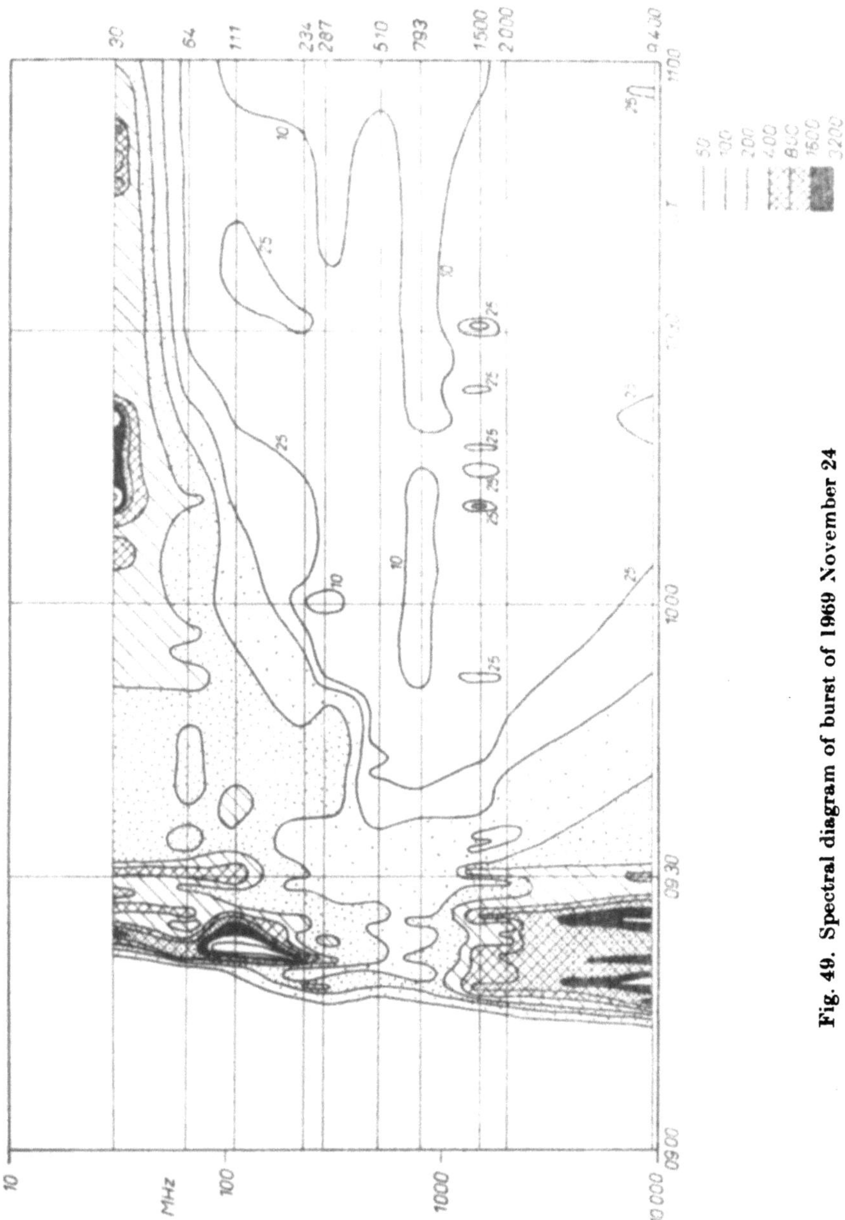

Fig. 49. Spectral diagram of burst of 1969 November 24

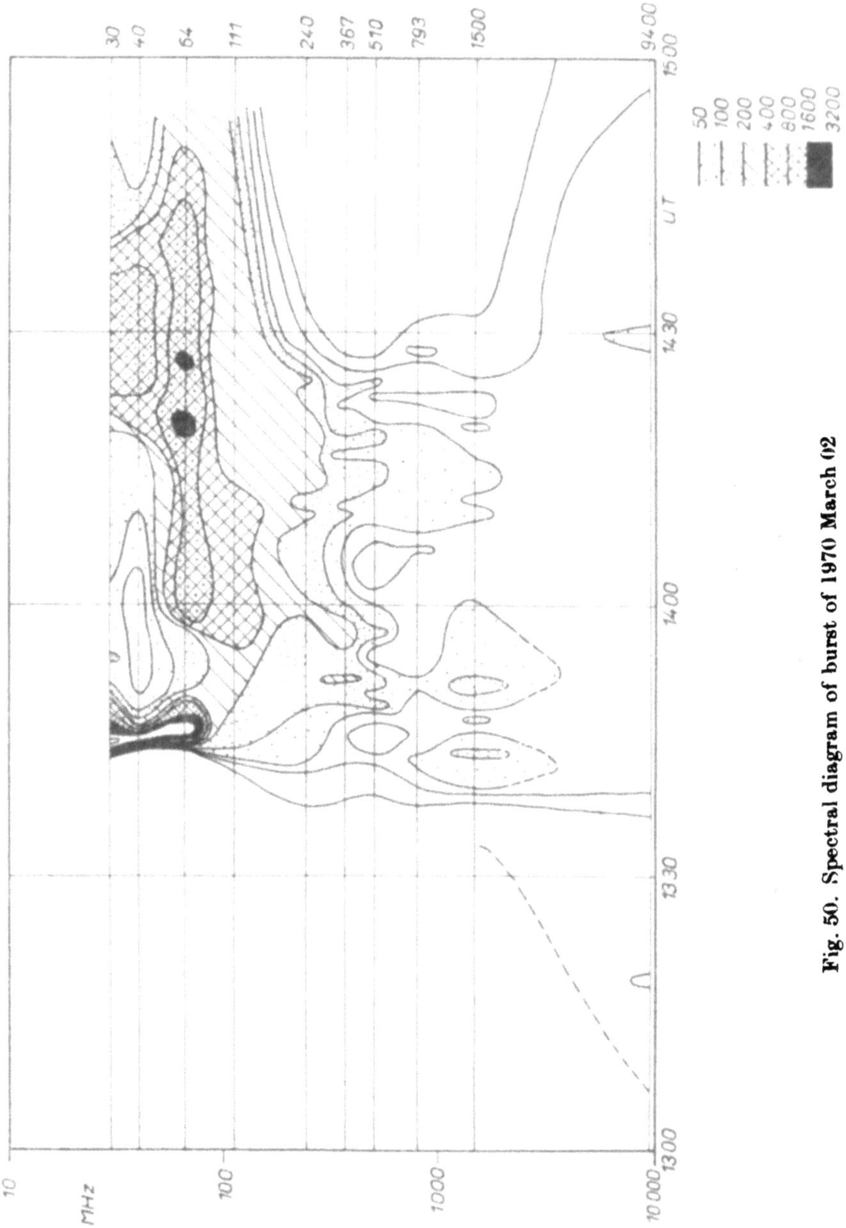

Fig. 50. Spectral diagram of burst of 1970 March 02

Index